NATURAL COMPUTING

ALSO BY DENNIS SHASHA AND CATHY LAZERE

Out of Their Minds:
The Lives and Discoveries of 15 Great Computer Scientists

ALSO BY DENNIS SHASHA

Puzzles for Programmers and Pros

The Puzzler's Elusion:
A Tale of Fraud, Pursuit, and the Art of Logic

Puzzling Adventures:
Tales of Strategy, Logic, and Mathematical Skill

Dr. Ecco, Mathematical Detective
(originally titled Codes, Puzzles, and Conspiracy)

Dr. Ecco's Cyberpuzzles:
36 Puzzles for Hackers and Other Mathematical Detectives

The Puzzling Adventures of Dr. Ecco

NATURAL COMPUTING

. . . .

DNA, Quantum Bits, and the Future of Smart Machines

DENNIS SHASHA
and CATHY LAZERE

W. W. NORTON & COMPANY
NEW YORK LONDON

Printed in the United States of America
First Edition

For information about permission to reproduce
selections from this book, write to Permissions,
W. W. Norton & Company, Inc.,
500 Fifth Avenue, New York, NY 10110

For information about special discounts for bulk
purchases, please contact W. W. Norton Special Sales at
specialsales@wwnorton.com or 800-233-4830

Manufacturing by Courier Westford
Book design by Lovedog Studio
Production manager: Devon Zahn

Library of Congress Cataloging-in-Publication Data

Shasha, Dennis Elliott.
Natural computing : DNA, quantum bits, and the future
of smart machines / Dennis Shasha and Cathy Lazere. — 1st ed.
p. cm.
Includes bibliographical references and index.
ISBN 978-0-393-33683-2 (pbk.)
1. Natural computation. 2. Artificial intelligence.
3. Computer scientists. I. Lazere, Cathy A. II. Title.
QA76.9.N37S33 2010
006.3—dc22

2010004807

W. W. Norton & Company, Inc.
500 Fifth Avenue, New York, N.Y. 10110
www.wwnorton.com

W. W. Norton & Company Ltd.
Castle House, 75/76 Wells Street, London W1T 3QT

1 2 3 4 5 6 7 8 9 0

To the troublemakers
who take science in new directions

CONTENTS

PREFACE

▪ ▪ ▪ ▪

When we began this book, we expected to talk to scientists who were solving problems on the frontier of what is possible in computing. We expected them to describe new computer architectures and a variety of new software techniques. The fifteen scientists featured in *Natural Computing* control spacecraft from millions of miles away, embed intelligence in smart bacteria, or build computers to run as fast as a million desktops combined. They work on the most challenging applications in science, engineering, and even finance. We expected this diversity, but we didn't expect the common vision that has emerged across all of these fields: *the future of computing is a synthesis with nature.*

There are three major strands to this vision.

First, biological thinking has inspired new ways to do digital computing. This hasn't happened yet in your word processor or in data centers, but it has occurred in applications that are pushing technology to its most interesting limits. For example, computers control spacecraft throughout flight and after landing. Manual repair of onboard hardware, once in flight, is often impossible;

but innovative spacecraft engineers propose designing machines that will repair themselves. If you're not in an engineering field, you may not appreciate what a change in thinking this approach represents. Instead of building a high-precision machine that can handle every possibility, designers construct a machine that will adapt itself to possibilities the designers cannot even imagine.

Although there is some debate about how to realize this new design philosophy, many of the scientists profiled in this book suggest that evolution or some form of learning should be involved. At first glance, evolution and learning may seem very different. Evolution works across many generations and organisms, while learning and adaptation occur in a single organism. Conceptually, however, they are very similar: try something and see how it works; then respond to feedback by trying something new that has a chance to be better. We associate learning with conscious improvement, but evolution occurs subconsciously, chemically, or even metabolically. For purposes of survival, however, learning and evolution have the same effect.

Many applications are based on evolutionary techniques. In *Natural Computing*, a surfer dude turned mathematical financier uses evolutionary algorithms to figure out when to buy or sell US Treasury bonds. A professor analyzes the safety of a missile defense system by checking how well safety procedures adapt to failure. In these instances, evolution, learning, and adaptation are linked.

Second, biological entities may replace silicon. Computing built on DNA or bacterial cells is nearly free (billions of bacteria need just a little sugar water to grow), and DNA computing also has a massively parallel capacity. One day, living bacteria rather than silicon electronics may compute inside our bodies or inside microfactories. Harnessing life in this way will require the development of an entirely new paradigm of computing. No longer will a human hand design computers to work reliably for several years, give them universally known names, and install them in air-conditioned splendor.

Computers made of bacteria or viruses come by the million, have no names, and are provincial—they communicate only with their neighbors. They also fail often. Getting them to do something useful is daunting and might seem impossible, but it must be possible because we already have what mathematicians call an existence proof: humans are composed of about 100 trillion cells and we are able to run, think, and love, even though none of our individual cells can do those things.

Third, new applications may require rethinking the underlying physics of computation. Simulating protein folding or breaking codes requires thousands of processors to coordinate their work. Because communication through switches and computer memory is slow compared with the speed of transistors, such a strategy works best when there is little communication between faraway processors. For that reason, the fastest digital machines on the planet are designed to move information as little as possible.

For example, a multi-millionaire inventor has embarked on the design of a machine to simulate the behavior of proteins. To do this, his hardware moves the information concerning each atom in the protein to very specific places within the active circuitry. At those places, information is combined with corresponding information about neighboring atoms—all without returning to the computer's slow memory. The payoff may be speeding up computation by a factor of 1,000, enough to accelerate a task that would normally require as much as a century to one that could be completed in less than 100 days.

Another designer, much less well funded, has constructed a computing device whose main computational element is a piece of foam attached to 25 wires. Measuring the electric current could offer a method to simulate the differential equations that characterize a whole host of "continuous" scientific problems, from the prediction of star trajectories in galaxies to the propagation of pigmentation on butterfly wings. This "extended analog computer" turns computing completely on its head. Instead of *calculating* an answer using 1s and 0s and arithmetic in a digital computer, it *measures* an answer.

You may find yourself feeling affection for these machines. Instead of the hard metal engines that do calculations or send e-mails or play chess, you will see machines that are more human—they repair themselves, attempt extremely hard problems, and sometimes make mistakes. These are computational devices that may one day set your broken bones, maintain the stability of bridges, or perhaps even help you breathe underwater.

In writing *Natural Computing*, we encountered a constellation of ideas that could change the world decisively for the better. Each chapter in this book describes a unique path to discovery. We hope you enjoy reading the stories of these risk-seeking adventurers as much as we enjoyed writing them.

Part I

ADAPTIVE COMPUTING

MODERN MANUFACTURING BEGAN WITH THE NOTION OF INTERchangeable parts, dating back to Johannes Gutenberg's movable type in the fifteenth century. By the eighteenth century, manufacturers had become more concerned about the precision of the parts. Eli Whitney's interchangeable musket parts had a precision of 1/30 inch (about 1 millimeter). Machine tolerances today are typically 10 micrometers (millionths of a meter), 100 times more precise than what Whitney could achieve. Optical tolerances are now measured in the nanometer range—a million times more precise than Whitney's. Designers now have the opportunity to build machines of exquisite precision for the task at hand.

Mainstream computer science is built on algorithms. An *algorithm* is a method that is guaranteed to produce a correct response for a large class of stimuli with a specified efficiency. Think of algorithms as recipes—to produce a particular dish, combine specified ingredients in a recommended order to obtain a desired result. For example, a "mergesort" algorithm puts items in order no matter what kinds of items are presented to it.

Although algorithms will always play a central role in computing, some problems are fundamentally not algorithmic. Consider the following problem: You are to survive in Antarctica and keep equipment operating at any temperature down to –60°C (–76°F). You know that your shelter and clothes may suffer any one of many possible mishaps. How will you and your equipment survive? An algorithmically oriented computer scientist would complain that the problem is ill posed. If the mishaps are great enough, it may not be possible to survive. But what if you had to design a solution that would work most of the time? You would need to build in adaptation or its cousin, evolution.

As early as 1954, the mathematician Nils Aall Barricelli, working at the Institute for Advanced Study at Princeton, simulated simple models of evolution using a computer. About fifteen years later, the German computer scientists Ingo Rechenberg and Hans-Paul Schwefel used evolution to improve designs for complex engineering problems.

In the natural world, evolution applies to organisms. In the computational world, evolution applies to designs. In both cases, evolution can lead to beautiful results *without the benefit of a conscious designer*. In 1975, John Holland of the University of Michigan wrote a landmark book, *Adaptation in Natural and Artificial Systems*, in which he showed the commonalities among the different approaches to evolutionary design and improved them through a uniform mathematical framework.

Holland's framework became the basis for modern genetic (sometimes called "evolutionary") algorithms. It consists of repeated applications of the following procedure:

1. Start with a population of possible designs (candidates).
2. Evaluate each one to give a "fitness" score, perhaps based on monetary cost or energy consumption.
3. Remember the design that receives the best fitness score.
4. Create a new population by selecting the fittest candidate designs and changing them slightly in a random way or changing them greatly by combining different designs together.

Suppose you are using this method to design a car. If a good design proposes a composite chassis with a six-cylinder engine and another good design proposes an aluminum chassis with an electric engine, the combined design might be a composite chassis with an electric engine. Parent designs beget children having some characteristics of each.

Although evolution can lead to better designs, small adaptations require less effort. For example, you can learn to ride a bicycle or juggle without evolving. Adaptations at that level may entail trial and error, but the organism doesn't need to change. **Rodney Brooks** does adaptation in motion. Since the 1980s, he has designed robots that move intelligently by adaptation. Since starting his pioneering work, Brooks has been inspired by insects, elephants, and geckos. In the process, he has redefined what it means for robots to be smart.

Suppose you are designing software for a robot that must navigate the surface of another planet. You don't know what the exact task will be. You do know the ground is rough. You also know the environment is extremely hostile, but you don't know the particulars. You are faced, in fact, not only with unknowns but with what **Glenn Reeves** of NASA's Jet Propulsion Laboratory calls "unknown unknowns"—unknowns you can't even characterize. It won't work to design a rover, send it, and then hope for the best. In fact, the current state of the art is to diagnose from afar—100 million miles away. To do that, Reeves has to design spacecraft instruments that act like chatty patients talking about their current condition and reporting when they feel better. When the instrument is sick, his team then has to send up electronic prosthetics in the form of *patches*, or small changes to the software. The device uses the patch in place of the erroneous code in the same way a patient would use a prosthetic limb in place of a non-functioning limb.

Adrian Stoica, who also works at the Jet Propulsion Laboratory, imagines future spacecraft that can heal themselves. Consider the challenges: day and nighttime temperatures on the surface of, say, Mars might vary from a frigid –133°C (–207°F) to a balmy +27°C (+80°F). A person facing temperature variations changes clothes. A circuit can't cover itself, but perhaps it can change how the electrons flow. Stoica has designed circuits that can "evolve" a solution on their own. Stoica dreams of equipment that will survive 100 years by evolutionary adaptation.

Some would argue that genetic algorithms do not deserve the good name of *algorithm*, because they don't guarantee efficiency and correctness. This is true. A genetic algorithm offers no guarantees, but it often arrives at surprisingly good designs for problems having no known algorithms. Using genetic algorithms, **Louis Qualls** designs custom nuclear power plants for extreme environments, including space. To get the specifications—how much power the plant should generate, how much it should weigh if powering a spacecraft, and so on—he talks to specialists, gets design preferences from each of them, and then tries to arrive at a design that embodies a compromise among those preferences while meeting the specifications. To do this, he must choose among trillions of design possibilities. By hand, he can explore only a fraction of the possibilities in his search for a good design. He also knows that he often returns to designs he has done previously because he has grown emotionally attached to them. By contrast, when he uses genetic algorithms to program his computer, the programs sometimes discover designs that he would never have considered but either cost less or perform better than his own. Further, if the specifications change (and they often do), he can set the computer off to find another design, still without any emotional attachment. It is quite likely that in the future, most complex engineering artifacts will be designed in this way.

Jake Loveless and **Amrut Bharambe** apply similar ideas to finance and use genetic algorithms to design rules to help them trade treasury bonds. While engineers use complex physical criteria to determine "fitness," Loveless and Bharambe use the most basic financial measures: high profit at relatively low risk. To determine whether a rule is good, they try it on historical data. The *search space* (the number of possible rules) is scarily large and they don't understand the rules that their genetic algorithms generate, but the method works.

Nancy Leveson works with nature too, but mostly human nature as it meshes with high technology. She considers the marvels of engineering—power plants, missile defense, and spacecraft—and tries to make them safe. She started her career in computing, but then took off for the jungles of New Guinea and pursued a serious interest in cognitive psychology before returning with a new perspective on computing. She believes that safety requires dependable adaptation—mistakes happen, but the system must compensate for them. The "system," in Leveson's view, does not stop at a convenient boundary like software specification. It extends all the way up the human management chain. When computing elements combine with people in life-critical situations, adaptability translates to multiple levels of feedback and modification. A failure at a lower level must be compensated for by a person or machine at a higher level. Leveson's approach tries to ensure that every level of a system detects problems at the level below and responds appropriately. Conceptually simple, this approach is being used to help prevent accidental missile launches and air traffic accidents.

■ ■ ■ ■

You're all assuming there's a logical representation inside the robot. What if there is no logical representation? I had been watching insects and how they do stuff . . . do they really have a three-dimensional rendering of the world around them—a computer graphics model inside that puny little head with 50,000 neurons? Is that what those neurons are doing?

— *Rodney Brooks*

Chapter 1

RODNEY BROOKS

■ ■ ■ ■

Animals Rule

WHEN ARTIFICIAL INTELLIGENCE (AI) WAS BORN IN THE 1950s, excelling at IQ tests or chess seemed to be a good indication of intelligence. After all, that's what schools measured. Since then, a slew of other definitions have been added to the mix, including emotional IQ and Howard Gardner's interpersonal and kinesthetic measures of intelligence. But if we define intelligence as the ability to survive in the world, we need to look at more fundamental skills. How is it that we can walk, recognize objects, and navigate around obstacles? You may say, "Animals can do that!" To which Rodney Brooks might respond, "Exactly!" In fact, robots might do better if they didn't copy humans in all aspects of our behavior. For example, who walks better over rough terrain—humans or insects? If you've ever seen insects scramble out of impossible holes, you might vote for the insects.

In a seminal paper from 1990, "Elephants Don't Play Chess," Rodney Brooks presented an evolutionary argument for the relative insignificance of human "higher" intelligence. Life arose on Earth 3.5 billion years ago, he noted; vertebrates and insects appeared during the last 10% of that time, approximately 450 million years ago. The great apes emerged in the last 0.5% of that period, about 18 million years ago. Agriculture was created only 19,000 years ago, 0.0005% of life's time on Earth. Expert knowledge has appeared only in the last few hundred years.

Computers are most successful at rapidly performing the skills learned in the last few hundred years of human history, perhaps because we are most conscious of those skills and they take the most conscious effort. But our unconscious acts pose a greater computing challenge. In the brief history of space travel, it has been easier to build a computer program to guide a spacecraft to Mars than to build a robot able to navigate over rough terrain with anything like the skill of a billy goat. Evolution required billions of years to arrive at the billy goat, but only a few million more to arrive at human intelligence. The factor of 1,000 in relative timescales should give us a certain humility before these "primitive" intelligences.

When Brooks wrote his "Elephants" paper, the field of artificial intelligence modeled intelligence as symbol manipulation. The scientific goal was to design sensor modules such as vision systems. These would abstract the world into symbols and pass the symbols into an intelligent core, a kind of electronic monarch. The monarch would manipulate the symbols and then instruct actuators (normally wheels) to move. In many ways, this stepped process mirrored the idealized hierarchy of a large corporation or the military—"brains" on top, eyes and limbs on the bottom. Brooks objected to this paradigm on philosophical as well as pragmatic grounds.

Brooks's philosophical objections may have originated in his

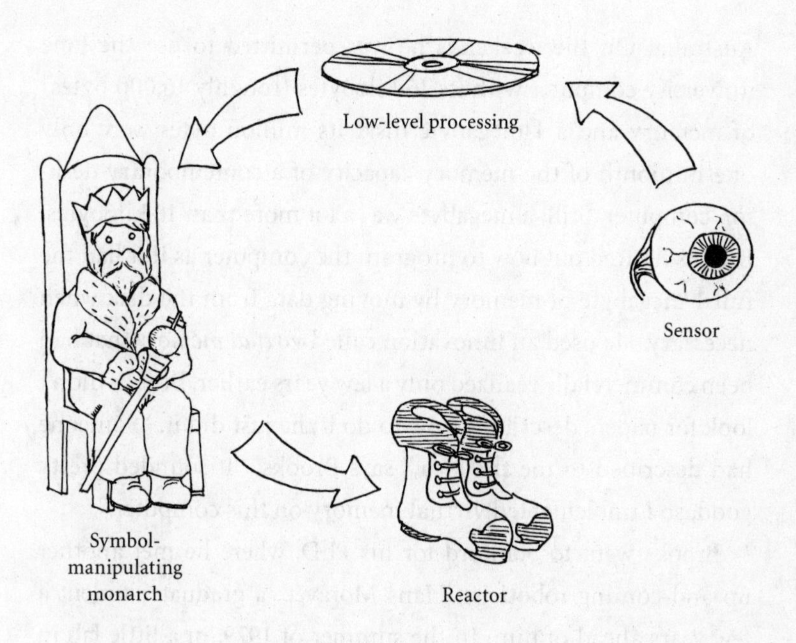

The monarch model. The monarch receives data from sensors, reasons about it, and then emits decisions that are carried out by reactors.

unlikely path to science and MIT, where he is now a professor at CSAIL, the Computer Science and Artificial Intelligence Laboratory. Brooks was born in 1954 in Adelaide, Australia. Though not exactly the outback, Adelaide was a long way from the centers of computer science research, but the remoteness may have been to Brooks's advantage. Nobody told him the right way to approach the field. At age eight, on his own, he began designing computers to play games. At twelve, he built a primitive computer out of old telephone relays to play tic-tac-toe. He resolved to pursue a career in game design.

In 1972, Brooks began studies at Flinders University of South

Australia. On the weekends he was permitted to use the lone university computer with its 16 kilobytes (roughly 16,000 bytes) of memory and a 1-megabyte disk. Its million bytes were only one-millionth of the memory capacity of a contemporary desktop computer. Still, a megabyte was a lot more than 16 kilobytes. Brooks figured out how to program the computer as if it had the full 1 megabyte of memory by moving data from the disk when necessary. He used an innovation called *virtual memory* that had been commercially realized only a few years earlier. Brooks didn't look for papers describing how to do it; he just did it. "Someone had described to me the idea," says Brooks. "It sounded pretty good, so I implemented virtual memory on this computer."

Brooks went to Stanford for his PhD, where he met another up-and-coming roboticist: Hans Moravec, a graduate student a few years ahead of him. In the summer of 1979, in a little lab in the hills behind the campus, Brooks helped Moravec with a robot known as Cart. At midnight each evening, when everyone else went home, Brooks and Moravec would set up Cart. The robot would move about 20 meters around the lab for the next 6 hours. The two young scientists wanted to create *stereo vision*—giving the robot depth perception by equipping it with slightly different images from each of two viewpoints. Stereo vision normally requires two cameras, but at that time cameras were expensive and they had only one. So they had to slide the one camera from side to side. "I was just a gofer to move furniture around, set things up, get things connected," says Brooks. The computer would have a look at the world and then compute for 15 minutes and move a meter. Then it would open its eyes, look, shut its eyes, and compute for another 15 minutes. It would move a meter blind, based on where it thought things were. "At the time I thought it was way too much computation," notes Brooks.

Brooks's doctoral thesis was on *machine vision*, a classic AI approach in which the camera would feed a computer an image. Brooks's program would then translate the visual scene into symbols to be processed by a hypothetical "intelligence"—the symbol-manipulating monarch. "The basic idea that nobody was questioning was that you've got a camera, you've got pixels, and you just change the pixels into a logical description of the world," Brooks says.

On vacation in Thailand in 1988, Brooks visited his first wife's family home, which stood on stilts by a river. No one spoke English, so Brooks sat by himself, watching insects. The more he watched, the more he began to question the symbolic AI paradigm. He just couldn't believe that insects were capable of forming logical descriptions inside "that puny little head with 50,000 neurons."

In 1990, Brooks's "Elephants" paper explained what he playfully called "nouvelle AI." His hypothesis was that an intelligent system had to have its representations grounded in the physical world. "The world is the best model of itself," as he put it. The world is up-to-date and contains all necessary details. This meant that Brooks's robots would dispense with the hierarchical structure of "classical" AI, with its symbolic representation of the world. Instead, nouvelle AI robots would possess a set of independently designed skills. Just as a human plays basketball and walks using the same limbs and eyes, a robot shares sensors and actuators for different skills. The skills, however, are independent—some of them, especially the highest-level ones, may fail without causing others to stop working.

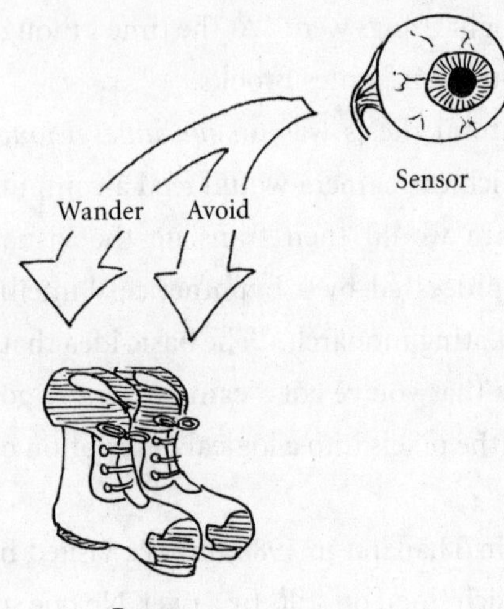

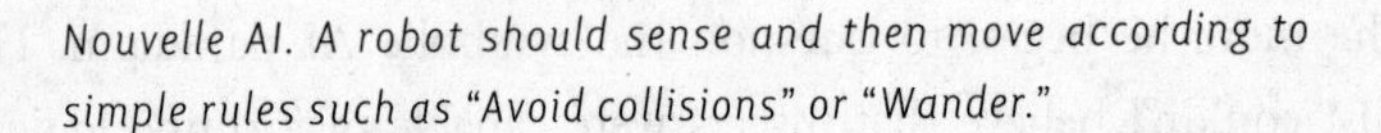
Nouvelle AI. A robot should sense and then move according to simple rules such as "Avoid collisions" or "Wander."

Embodying this philosophy, the Brooks lab at MIT had built an early robot called Allen in 1985, which Brooks puckishly had named after Allen Newell, one of the early proponents of symbolic AI. Allen had three skills: avoid collisions, wander around randomly, and go to distant objects. "Allen would happily sit in the middle of a room until approached, and then scurry away, avoiding collisions as it went," remembers Brooks. "The internal representation was that every sonar return represented a repulsive force."

At the lowest level, Allen acted like a frightened mouse, following a primal rule: avoid hitting or being hit. What would keep Allen from hiding in a corner? Every 10 seconds, Allen would be told to wander randomly. Note that wandering also requires moving wheels, in accordance with the Brooks strategy of having different skills use the same actuators. The collision avoidance skill takes precedence over wandering.

In its third function, the robot used its sonar to look for distant places and try to go to them. It would measure distance using an odometer. Like a runner trying to complete a mountain race while avoiding a slip off the edge, the robot combined goal seeking with underlying survival skills. It seemed almost too easy to be research. "I argue for simplicity," says Brooks. "Get away from hairy equations." He contrasts that viewpoint with the belief of some of his colleagues and critics, who think the hairier the better. To Brooks, if you need to explain something with a lot of convoluted mathematics, the solution will be "pretty unstable." "I'm interested in building something that can't fail to work," he says.

Rolling robots were one thing. What about walking robots for scrambling over rough terrain? All that thinking in Thailand would be put into play. Working with Colin Angle and Grinnell Moore, a high school student, Brooks built a six-legged walking robot named Genghis. "There weren't many walking robots at the time. Everyone else's walking machine was big and fragile," remembers Brooks.

Brooks had been looking at high-speed videos of insects running. "They fall all the time and hit their metaphorical chin," he says. Because the strength-to-weight ratio is greater at smaller sizes, they can afford to be really bad walkers. By contrast, a filly at the Kentucky Derby can't afford to fall—she might break a leg and suffer a career-ending injury.

Brooks found a natural application for Genghis at the Jet Propulsion Laboratory (JPL). He started attending meetings as they were talking about new Mars missions. JPL was promoting a

Brooks's walking robot Genghis, a multi-legged, low-lying device that could afford to fall.

1,000-kilogram robot called Robbie with a big arm at the front. It moved a centimeter a minute. JPL estimated that a mission to send Robbie to Mars would cost about $12 billion. Brooks knew the required funding would never come through. At a meeting, he suggested sending a small robot instead of a big one. "I remember one of the lifers at JPL who was an instrument builder saying [as he tells the story, Brooks morphs his Aussie accent into a southern drawl], 'A scientist waits 15 or 20 years for a mission, he doesn't want a little bitty instrument—he wants a *big* instrument. You can't send a little robot—you've got to send a *big* robot!'"

Brooks proposed a compromise. Instead of sending a single 1,000-kilogram robot, send a hundred 1-kilogram robots. That would cut the total mass down by a factor of 10 and spread the risk. "If you send a 1,000-kilogram robot, you have to be very careful about what you do," says Brooks. "If you screw up, you've lost your $12 billion investment. If you've got 100 of them and you lose one—big deal."

Brooks found another benefit in the "swarm of robots" idea: mass production. Building 100 copies of a small robot would be cheaper than building a single 1,000-kilogram robot. But then what do you do with 100 robots once they get to Mars? "If you have a single 1,000-kilogram robot, you want to control it at all times. If you have 100 [smaller robots], you can't possibly control them; they have to be autonomous. So get over it—make them autonomous from day one and let them out of your control."

In 1989, Brooks wrote a paper for the British Interplanetary Society embodying his in-your-face plan. The title of that paper—"Fast, Cheap, and Out of Control"—was later used in 1997 for a documentary by Erroll Morris in which Brooks appeared, along with a topiary artist, a lion tamer, and the world's foremost expert on naked moles. The paper resonated with a NASA scientist named Donna Shirley. Shirley arranged for Colin Angle, an undergraduate at the time, to spend the summer of 1989 at the Jet Propulsion Laboratory. There Angle built a little, half-kilogram robot called Tooth that could do many of the things the 1,000-kilogram JPL robot Robbie could do. Rajiv Desau and Dave Miller at JPL used the Tooth code to build Rocky I and Rocky II, six-wheeled rovers. The current Mars rovers are based on that design.

Inspired by that success, Brooks and the newly minted graduate Colin Angle decided to start a company for lunar and Mars exploration. "We were going to send pairs of robots to the moon with advertising decals on them," says Brooks. Eventually, Brooks teamed up with David Scott, commander of *Apollo 15.* Through Scott, Brooks became connected in 1992 to the Strategic Defense Initiative (SDI), fondly labeled "Star Wars" by its opponents and, under Clinton, renamed the Ballistic Missile Defense Organization. The plan was to exploit bureaucratic rivalry. The developers of ballistic missile defense had spacecraft called Brilliant Pebbles. The idea was that these spacecraft would locate rocket plumes of ballistic missiles in their launch phase and then ram them with non-nuclear mini-missiles. But by 1992, the Soviet threat had disappeared, so the Ballistic Missile Defense Organization needed to show off its technology in other ways.

A new plan was proposed to put a Brilliant Pebble on a vehicle called Clementine that would orbit the moon. The missile detection instruments of the pebble would be replaced with Brooks's walking robot called Grendel. The pebble would land on the moon, and the robot would do some environmental tests. In 1993, a test run was conducted at Edwards Air Force Base. It looked as though it could work. Interagency turf wars quickly interfered, however. NASA protested that extraterrestrial exploration was its job, so the idea was scrapped. Brooks's Rocky VI, however, was sent to Mars. NASA seemed on board with the Brooks philosophy. "They said we're going to do it faster, cheaper, better. We're not going to do it the old way. We're going to do it smarter," Brooks recalls. Then the next mission to Mars failed. "Faster, cheaper, better got canned."

Since then, Brooks's company, iRobot, has been making autonomous vacuum cleaners called Roombas for household use. It also manufactures bomb-disarming robots for the military. The bioengineer Robert Full has collaborated with iRobot on robot construction and has revisited the insect inspiration. After studying the movements of cockroaches and similar crawling insects with colleagues, Full determined that the legs of such creatures could be modeled as articulated pogo sticks. Designing moving robots this way enables them to navigate over difficult obstacles and to move in and out of craters without a vision system and with minimal computation. Springy limbs have eliminated the need for thought in locomotion. Simplicity wins.

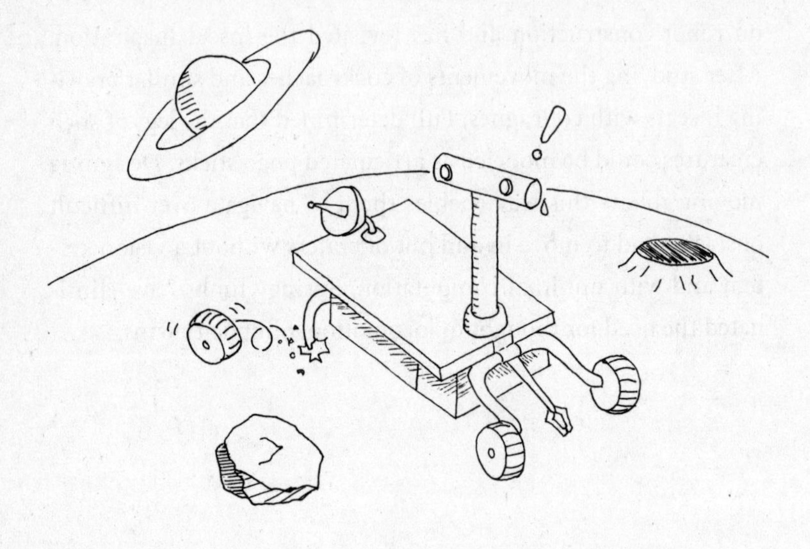

■ ■ ■ ■

When you're working with equipment on Earth, how you diagnose it and how you repair it are based on the fact that quite literally you can walk up to it and touch it. The minute it leaves the Earth, none of those things exist. You can't touch it—you can only observe it by what it tells you. You can't go fix it. So you have to deal with the fact that if something is going to break, you either have to give up, or you have to work around it.

— *Glenn Reeves*

If you send out a spacecraft that will reach its target in 100 years, you can't afford to use electronics that will fail within 10 years. You can try massive redundancy, but then you might make a spacecraft that can't fly. The goal is to see if these systems can become long life survivable all by themselves.

— *Adrian Stoica*

Chapter 2

GLENN REEVES and ADRIAN STOICA

. . . .

Design for a Faraway Planet

IMAGINE YOU HAVE TO TROUBLESHOOT A MULTI-MILLION-DOLLAR production system at a customer site. All eyes are on you—the customer's, your boss's, and, if you're in charge, your colleagues'. To analyze the problem, you test individual software and hardware components, think, change the data, and run the software. They're watching you. You think some more, run the existing data on different hardware, think as clearly as possible. They're still watching. You think of a fix that might work. You try it . . .

Now imagine that your equipment is unique to your application. Not only are your customers looking at you, but so is the world press. The spacecraft is several minutes away at the speed of light. In the best case, weather and power permitting, it can communicate with you at 1,200 bits per second (a DSL line communicates 5,000 times faster). You can't change the hardware or even physically look at it.

Next, imagine that the spacecraft's mission is to a distant

planet, so far away that it takes 100 years to get there. Once there, the spacecraft is subject to extremes in heat and cold. If a part breaks, there's no FedEx to replace it. Now suppose that somehow the spacecraft can fix itself. Welcome to the challenges faced by Glenn Reeves and Adrian Stoica of the Jet Propulsion Laboratory.

Fix and Learn

Born in 1960, Glenn Reeves grew up in South Pasadena. His father worked as a civil engineer and manager for Caltrans, the California State department in charge of freeways. As a young boy, Reeves was interested in outer space. At his fifth birthday party, cardboard cutouts of the moon, the planets, and stars hung from the trees in the backyard. Other more grown-up dreamers lived in the neighborhood—Pasadena is the home of the Jet Propulsion Laboratory (JPL).

Set against the San Gabriel Mountains, JPL sprawls over almost two hundred acres. It was founded in the 1930s by the California Institute of Technology (Caltech). In the wake of *Sputnik*, JPL designed the first American satellite, *Explorer I*, which was launched in 1958, shortly before Reeves was born. JPL is essentially a laboratory for NASA; its employees design spacecraft and robotic missions to explore the moon and the planets.

"We would drive by JPL when I was young," remembers Reeves. "My mom would point as we were driving by and say to me, 'You know you're going to work there someday.' I didn't even know what the place was." In high school, Reeves pursued science, but only halfheartedly. He liked physics but didn't excel at it. He took a math self-study program and fell behind. His main passion was working as an auto mechanic. "I was far more interested in having a job and making some money than preparing for a career," he admits.

At high school graduation, Reeves knew he should go to college—his parents had insisted on that—but where to go and what to study escaped him. He planned to go to the University of California at Santa Barbara, but then he became romantically involved with a local woman and decided to attend Cal Poly at Pomona instead. There he took introductory courses in hotel and restaurant management, accounting, and information systems. It was his first exposure to the computers of the day, which required writing instructions on Fortran punch cards. "That little puzzle of how to do that very simple programming had me sufficiently intrigued to say, well, maybe I'd like to do this," Reeves recounts.

Reeves majored in computer science and continued living in South Pasadena, commuting every day to Pomona. A chance meeting changed his life. On his way to school one day, he noticed one of his professors, John Rohr, sitting at a bus stop. Reeves was en route to Rohr's class and realized that if Rohr kept waiting for the bus, the class wouldn't happen. So he offered Rohr a ride. In the course of their conversation during that drive, Rohr talked about his work on the software for deep-space missions at JPL.

Rohr became Reeves's advisor and arranged a summer job for Reeves at JPL. That short-term job led to a full-time position after Reeves graduated in 1983. He was assigned as a programmer for the spacecraft test equipment of the *Magellan* spacecraft. Named after the Portuguese explorer Ferdinand Magellan, the spacecraft's mission was to map the surface of Venus.

Spacecraft test equipment of that era had two roles. First, it had to act like the ground control equipment, the electronics that communicates with the spacecraft. Second, it had to act like the spacecraft itself, including the parts that hadn't been built yet. With these functions, the test equipment enabled the development team to build, test, and simulate flying the spacecraft long before the final assembly was completed.

Reeves's boss, Robert Anderson, broke with tradition by replacing the mainframe-based spacecraft test equipment with many interconnected single-board computers. In retrospect, the advent of microprocessors made this a logical choice. Microprocessors simplified the architecture: many independent boards reduced the likelihood of a full system failure and eliminated the single mainframe as a potential bottleneck. Reeves considers Anderson to be his most influential mentor: "He taught me that sometimes you have to step back and start with a blank sheet of paper. There's a point where capitalizing on your legacy and heritage is no longer the right path forward anymore."

After the *Magellan* project, Reeves did a brief stint outside JPL. As in many organizations, leaving and then returning is one way to increase pay and responsibility. In 1991, he came back as the leader of the team developing the software for the test equipment for the *Cassini* spacecraft. *Cassini*'s mission was to explore Saturn's rings and moons. Because *Cassini* had many more instruments than *Magellan*, Reeves extended Anderson's distributed architecture to a greater number of interfaces and simulation requirements. Each device on the spacecraft had a corresponding dedicated computer, software, and specialized hardware interface. About 20 computers were coordinated and time-synchronized. They logged data, automated testing for consistency, and provided connections to human ground operators.

Cassini was successfully launched in October 1997. In the

course of its mission, *Cassini* entered Saturn's orbit on June 30, 2004. In January 2005 a major component, the *Huygens* probe, built by the European Space Agency, dove toward one of Saturn's moons, Titan, to report on the composition of the atmosphere before it landed on the surface.

In the 1990s, NASA also initiated a series of robotic missions to Mars. It may come as a surprise to us wingless bipeds that land travel is far more difficult than flying. The challenge begins with the mechanical puzzle of landing without crushing the equipment. Once on the surface, the hardware must cope with dust, extreme temperature variations, and the difficulty of phoning home. These robotic missions clearly required a new view of design. Because of his success with prior projects, Reeves was selected to lead the team working on spacecraft software development ("flight software") for the *Mars Pathfinder* mission.

The new challenges must have seemed daunting. The spacecraft had to perform three independent missions: launch and travel to Mars, land successfully, and then act as a weather station and communication station for the *Sojourner* rover. The spacecraft had to accomplish all of this after surviving an almost crash landing on Mars and coping with limited power while being bombarded with powerful radiation.

After launch, no human can touch a spacecraft's equipment or probe it physically. Even electronic communication is fragile. "If the antenna pointed toward Earth is off a little bit, you won't hear the spacecraft, because it's transmitting with a power of what amounts to a 100-watt bulb. That's compounded by how slow it is—how many minutes it takes for a round-trip," explains Reeves. With all of these potential problems, things tend to go wrong. Fully 50% of the Mars missions have suffered failures, and most of those failures have resulted in the catastrophic loss of the spacecraft. Many spacecraft are built with redundant hardware, and careful thought is given to the problems and failures that might develop. But not everything can be anticipated. Not one of the Mars missions has been error-free.

There are two philosophies about what to do when problems arise: (1) abandon the mission or (2) try a workaround. Reeves, the former auto mechanic, believes in anticipating problems as much as possible while building in tools to work around problems he doesn't anticipate—the "unknown unknowns," as he calls them. This philosophy comes in handy when you have to salvage a multi-million-dollar spacecraft. The fix-it philosophy also ensures that lessons learned from the problems become part of the engineering knowledge base. "If you give up, then you'll never learn anything," says Reeves. "You won't know what to fix in the future because you won't know what went wrong."

As most engineers know, there are good failures and bad ones. If there is a single root cause and it's something you could fix in the future, that's a good failure. If you can't find the root cause, then fantasies take over and you don't know what to do next. After all, the spectrum of possible causes is enormous—workmanship problems on one-of-a-kind devices, the harsh environment, dust, radiation, and aliens, to name just a few. "We haven't flown enough spacecraft to really understand the space environment. We have some knowledge, but we're still somewhat guessing," admits Reeves.

So far, the failures have been good ones; indeed, all of the failures to date could have happened on Earth. To understand both the errors and their solutions, we need to understand a bit about the debugging setup. For each spacecraft it sends into space, JPL builds what amounts to a duplicate spacecraft on Earth. That way, activities can be practiced and commands can be verified. If the behavior of the spacecraft is abnormal, the Earth-bound mechanics have a platform for diagnosis and resolution.

Spacecraft are designed to send data and status information from time to time. If, for example, a basic "system OK" status message arrives but scientific data doesn't, there might be a problem with a scientific instrument but not with the whole device. Ground control can take an active debugging role by issuing commands that change the behavior of the spacecraft, notably by uploading software corrections or by restarting various components. Just as we often "fix" a problem on a personal computer by restarting the machine, we can do the same with spacecraft computers. As on Earth, however, restarting is sometimes not enough.

The *Mars Pathfinder* consisted of two vehicles: the *Sojourner* rover (a mobile robot designed to do experiments on the planet's surface) and the lander that provided the communication station, the imaging, and the weather station capabilities. Reeves worked on the lander. His job was to develop the software that took the spacecraft from Earth to the Martian orbit, performed the landing, and then executed activities on the surface. The rover operated independently; communication would come through the lander and be relayed to Earth.

To world acclaim on July 4, 1997, the *Pathfinder* landed on the surface of Mars, bouncing happily on its airbags. It started to collect scientific data and perform its mission. But a few days later, the lander's computer began repeatedly restarting itself, stopping all useful work. Reeves was asked to find out why. "If a computer is sitting in front of you, you hit Ctrl-Alt-Delete and you reboot.

The basic theory is the same in space," says Reeves. "The hard part is we need to know what the vehicle was doing just prior to the reset."

Fortunately, Reeves's group had designed the ability to trace what the computer was doing while it executed. They had a record of all messages sent, which significant events had occurred in the software, and, above all, which tasks were being executed and when. The debugging team* determined that a low-priority task was blocking a high-priority task. It was as if a stream of taxis had occupied an intersection and blocked an ambulance that was rushing toward a hospital.

The lander's operating system, developed by a company called Wind River, allowed higher-priority tasks to stop lower-priority tasks from running, except when the lower-priority task held a "lock" on a particular resource that the higher-priority task needed. In this case, the resource was a queue used to pass messages containing scientific data to one of the science-processing tasks. "Our first question was, Why should the low-priority task hold the lock for so long? This is where having a good test bed on Earth came in handy," notes Reeves. The team could use exactly the same software on the test bed as on the flight vehicle. Over the course of several days, they were able to cause the same problem to occur. The problem then became obvious—it was a priority inversion situation (see the box "Priority Inversion").

* The debugging team consisted of Rick Achatz, Dave Cummings, Kim Gostelow, Don Meyer, Karl Schneider, Dave Smyth, Steve Stolper, Greg Welz, and Pam Yoshioka at JPL, along with Mike Deliman, Brian Lazara, and Lisa Stanley at Wind River, the vendor of the VxWorks operating system that was used on the *Mars Pathfinder.*

Priority Inversion: Three Tasks and a Lock

Priority inversion occurs because of a turf war between three tasks and a "lock." The low-priority task (L) acquires the lock and begins to use a resource. The high-priority task (H) stops L from running the resource, and then H runs itself. But H stops when it needs the lock and wants to use the resource. At this point, a middle-priority task (M) may begin to run. M doesn't need the lock, so it executes without allowing L to complete and release the lock. In effect, M prevents H from executing, thus "inverting" priority.*

The best-known solution to priority inversion, proposed by Lui Sha, Ragunathan Rajkumar, and John P. Lehoczky of Carnegie Mellon University, is to raise the priority of L to the same level as H when H finds itself in need of the lock that L has. That is, L "inherits" the priority of H. The net effect is to give priority of L over M until L releases the lock.

■ ■ ■ ■

* Aidan Daly, our illustrator, suggests the following good consequence of priority inversion from Tolkien's fictional "Middle Earth":

- The Hobbits (low priority) have a lock on the One Ring
- Sauron (high priority) cannot be resurrected and rule without the One Ring
- Thus, Men (medium priority) can rule in peace.

Fixing this problem seemed simple—just change the software. Wind River's software offered an easy way to make the change. But how should this be done from millions of miles away? On your desktop, installing a fix can mean installing a new CD. The entire transport of a new piece of software consists of taking the media, putting it into the computer, and then running the program it has installed. In the case of the *Sojourner* rover, Reeves and his group had to change software across 100 million miles, making the transport part of the fix quite challenging. The speed of transmission from Earth to the spacecraft is 1,200 bits per second or less. Reeves wanted to avoid sending all the software, because most of it hadn't changed. Instead, he sent only the differences between the existing code and the new code.

Changing software is a little like heart surgery. You want everything in place before you wake the patient. In the case of the *Pathfinder*, Reeves's team had anticipated the need to change the software. Their design allowed room for two copies of the software: an active one and an inactive one. Thus, they were able to merge the changes sent from the ground with a copy of the already-loaded software and put it in the inactive storage. They then checked that the software had been correctly received. Only after that did they make that copy active.

Why hadn't this problem been found in testing? A quick response would be to cite time pressure, which had encouraged the testers simply to reset the computer whenever this problem arose. A more thoughtful response involves the nature of organizational learning. Reeves describes it this way: "When you build a spacecraft, you actually learn how to operate it. As you get better at operating it, you commit it to doing more and more things because you better understand how it works and what you can accomplish. In the early parts of testing, you're actually fairly timid. Even though you try to test it as you're going to use it, sometimes you don't drive it hard enough." By the time the spacecraft landed, the scientists were more and more confident; they had been using the new instruments extensively.

Some of Reeves's colleagues summarize this lesson as "test what you fly; fly what you test." That is, the spacecraft shouldn't be asked to do anything more than it did during the test stage. In their view, problems arise from violating this principle. The principle makes sense as an ideal, but the harsh environment and, in this case, increasing confidence make the "fly what you test" goal unattainable in practice. An alternative lesson is "prepare for the unexpected."

In 2000, Reeves was assigned to the Mars Exploration Rover mission. He quickly understood another lesson: on Mars, hardware doesn't last long. Temperatures can vacillate between almost 0°C (32°F) during the day and –100°C (-148°F) at night. The batteries can keep the equipment as warm as –40°C (-40°F) at night, but such extreme variations cause fatigue in the hardware, break solder drums, and eventually cause electrical problems. Vehicles with a 90-day estimated life span can't afford to lose 15 days while bugs are fixed. If problems arose, Reeves and his team would have to resolve them fast.

What's more, Reeves now had to write the communication software for *two* mobile robots—two, in order to allow for failure. Mobility introduces a choice: exercise control from the ground or give the robot autonomy. In computer science terms, this is a choice between high-level and low-level control. Does ground control give a series of commands such as "Move the left front wheel six times and the right front wheel five times"? Or is the command something like this: "Drive the vehicle from here to there, taking care to avoid obstacles"? Or is it as simple as "Find something cool!"? The first option would prevent a sufficient amount of science from being accomplished.

An additional problem in communicating with something on the surface of Mars is that the red planet is rotating. When a spacecraft is in space, Earth rotates and one of the ground stations will always be able to see and talk to the craft, so there is 24-hour coverage. As Mars rotates, however, the rover moves with the planet and ends up pointing away from Earth, where it can no longer communicate with any ground station.

The "find something cool" option requires a high level of scientific knowledge. Much of robotic exploration occurs in response to ongoing discoveries. If the robot discovers an unusual chemical in a rock, ground control will instruct the robot to perform more experiments. Currently, the scientific team determines what is cool. Reeves believes this will change. As exploration reaches beyond 50 hours of light time, control on Earth won't be able to tell the lander to go look at a particular object. "The robots will need much more autonomy," predicts Reeves. Computational devices will require a great deal more scientific know-how and the ability to repair themselves. That's what Reeves's colleague at JPL, Adrian Stoica, is exploring.

Adapt and Prosper

Consider the problem of designing for a 100-year mission. High-energy particles bombard a spacecraft's sensitive electronic innards. Extreme temperatures cause mechanical wear and tear. Radiation and temperature variations like these currently demand active shielding—sort of like high-tech thermos bottles. But these conventional solutions require more weight and energy—two resources that are notably scarce in space missions.

Of course, humans can't tolerate such extremes either. Like our primate cousins, we are optimized for about 22°C (70°F). As a species, we have adapted to a range from roughly –40°C to +40°C (–40°F to +104°F) by wearing clothes or avoiding intense sunlight. By changing their own circuit connections and voltages, electronics may be able to adapt and avoid the heavy-thermos-bottle syndrome. Just as a child knows to put on a jacket in response to a chill wind, it's possible that a circuit could adjust itself when it notices that there has been a change in its environment or that it has failed at a particular task. This is Adrian Stoica's radical research proposal.

Stoica was born in 1962 in Iasi, capital of the Romanian principality of Moldova (in English, Moldavia). Moldova united with Transylvania and Wallachia to form modern Romania. But after World War II, part of Moldova was forced into the Soviet sphere. Long considered Romania's cultural capital, Iasi is the home of its oldest university, where Stoica's father and brother have taught economics. His mother worked as an accountant at the National Theater.

The Iasi schools supported rigorous programs in math and science. "It was very tough on children," says Stoica. But he is grateful for the training he received there, which he believes gave him an advantage over kids in other parts of the country. Romanian high school science was built around the best American and

Russian texts in math and physics. Students were able to access Western works in translation, such as Feynman's physics series and the Berkeley courses. By the time he reached the Technical University of Iasi in 1981, Stoica was well prepared. He soon became passionate about electronics and computing. As a freshman, he began programming with punch cards, and by graduation he was able to build microprocessor systems on his own.

In those pre-Internet days, Romanian libraries had little advanced technical literature. Although books on math and science were available, books from the West on technology were scarce; students had to photocopy and share whatever was available. Electrical components were cadged from foreign students attending Romanian universities. "We were buying stuff on the black market for our university projects," says Stoica.

In his free time Stoica read science fiction. Among his favorite writers were Isaac Asimov, Stanislav Lem, and other European science fiction authors. He found them inspirational. "Science fiction opens your mind and makes you dream. Later on in life I heard a quote approximately saying, 'If you can see it in your mind, you can do it in real life,'" Stoica recalls. Encouraged by science fiction and movies like *Blade Runner*, Stoica imagined himself making humanoid robots in a small lab or corporation. He graduated with an MS in electrical engineering from the Technical University of Iasi in 1986 and went to work for a company that made sensors and measurement equipment. He remembers telling his colleagues, "Guys, one day I will be working for NASA." Stoica admits that everyone laughed. "But I was serious. I can't say whether it was a premonition or simply a way to express my dream so it would come true that one day I would be able to contribute by working at the most advanced place in technology."

Dreams of NASA were delayed while Stoica pursued an academic career in Australia. At Victoria University in Melbourne, his doctoral work involved teaching robots to move by observing a human demonstrator. As he was completing his thesis in 1996, Stoica won the "green card lottery"—the United States' Diversity Visa program, which enabled him to immigrate to the United States. To satisfy immigration requirements, he dropped everything and moved to the States to work at the Jet Propulsion Laboratory and his dream boss, NASA. He has been there ever since.

Stoica's first job was to build electronics that could function on the surfaces of other planets or in extreme environments on Earth, such as volcanoes, nuclear reactors, and drill holes for oil wells. The thermos-bottle approach was to shield the electronics by isolating them from the environment. Stoica thought of a second possibility: replacing broken parts when the environment changes by having the electronics reconfigure themselves.

Stoica noted that six missions—*Voyager 1* and *2*, *Viking 1* and *2*, *Galileo*, and *Magellan*—all had problems following launch. Spare parts played a critical role in saving these missions. Spare or redundant parts clearly help, but they take up space and weight, which means less room for scientific instruments. More weight also means more expense (roughly $10,000 per pound). Equipping spacecraft with enough spare parts for any eventuality could easily make a mission unaffordable, because designers can never know in advance how many are enough. Instead, Stoica wanted to achieve reliability by building self-reconfigurable components. This approach offers more flexibility and versatility for the electronics but requires precisely locating the fault inside the component and working around the failure.

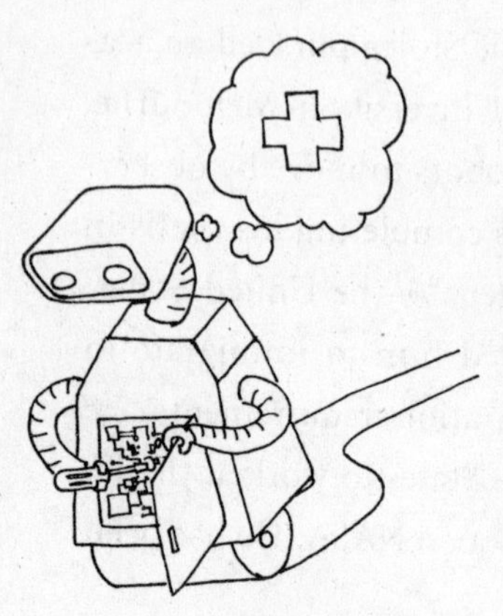

The human body offers a good analogy. Cuts normally heal (by local skin regrowth) in a few days; broken arms, in a few weeks. But an amputated limb requires a prosthetic. If a spacecraft could be designed so that minor failures would be repaired locally and severe failures would be repaired by replacement with a spare part, then the spacecraft might survive for a long time with just a few spare parts.

Stoica thought he could achieve such adaptability by using evolution. When an environment is static and benign, specialization wins the day; when things can change from one day to the next, versatility is essential to survival. This is true for bacteria assaulted by an antibiotic, for plants under environmental stress, and for people in a dynamic technological field.

Stoica heard about the work of inventor (and consulting professor at Stanford) John Koza, who had taken the genetic algorithm idea of his advisor, John Holland, and applied it to circuits. In Koza's framework, a genetic algorithm would start with small collections of resistors, transistors, and capacitors. Given a problem's specification, such as creating a stereo tuner, it would then develop an efficient circuit to solve the problem. In a process inspired by biological evolution, the algorithm selects the circuits from a population that best meets specifications and combines their genetic codes to form new offspring circuits, again selecting the fittest ones (in this example, those that tune the music best), and so on.

Stoica's goal differed from Koza's, however. Koza used evolutionary principles to design a circuit that solved a problem. Stoica designed and built flexible analog circuits that would use a genetic approach to reconfigure the circuit *after it was built* to adapt to circumstances. "Our thought was to use evolution in real time," says Stoica. Stoica sought a stand-alone, self-evolving chip. Under radiation or temperature changes, certain characteristics of the devices would change. A redesign might find a different optimal circuit topology or a different set of voltage biases in order to achieve optimal performance in a new environment.

The hardware that Stoica used is called a *field-programmable transistor array*. It is a two-dimensional, Manhattan-like grid of transistors and other components interconnected by other transistors acting as switches. The switches can be either closed (binary value 1), allowing current flow; or open (binary value 0), preventing current flow. Switches are under the binary control of the bit string (sequence of 0s and 1s) that defines the "genetic code" of the circuit. Through an evolutionary search, various combinations of bit strings are tried, thus helping the circuit improve its performance. Stoica has been able to show that his circuits can adapt even under temperature variations ranging from –180°C to +120°C (–292°F to +248°F).

Stoica envisions a hierarchy of adaptation. There will be some small closed loops where the system is adaptive at the cell level (like skin repairing a cut). And then if there are more cells, groups of cells will show some self-organization. The result will be a higher level of coordination and evolution. Adaptive parts enhance survivability, but could they pose a danger and get out of control? Stoica imagines one negative possibility straight out of the science fiction he loved as a teenager: "If at some point we combine these artificial systems with living systems or allow them to replicate, then shutting them down if they go rogue is not as simple as unplugging the power. They may go out of control because of numbers and replication." He maintains, however, that the positives still outweigh the negatives.

In formulating advanced concepts for space, Stoica has played with the idea of *terraforming* on other planets—creating human-friendly habitats on Mars, for example. "It's a mixture of the living and the artificial with a functional purpose." Maybe when humans arrive at a terraformed Mars, robots will form a welcoming committee. They might offer champagne. Cheers?

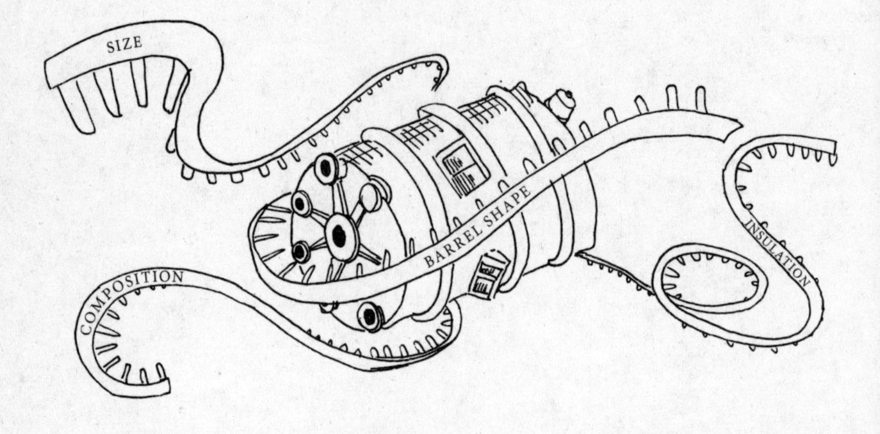

■ ■ ■ ■

Genetic algorithms are a tool that might tell me something about the system I want to design that I didn't think about because of my own biases.

— *Louis Qualls*

Chapter 3

LOUIS QUALLS

■ ■ ■ ■

Putting Evolution on the Design Team

DURING WORLD WAR II, A SECRET CITY AROSE IN THE HILLS OF Tennessee. Oak Ridge National Laboratory started as part of the Manhattan Project with a mandate to produce and purify plutonium for the atomic bomb. Since then, the laboratory has diversified into energy, systems biology, and materials science research. It has maintained a world-class leadership role in nuclear physics and engineering. In a time of budget crunches, the US Department of Energy supports the scientists at Oak Ridge with generous funding and large-scale engineering challenges.

Lou Qualls, a senior researcher and nuclear engineer at Oak Ridge, describes himself as a *system integrator.* He puts different subcomponents together in order to realize complicated engineering designs for nuclear reactor systems. System integration mixes psychology and engineering. It involves persuading subcomponent specialists that their ideal design will have to change in order to fit into an overall framework. Artists tend to fall in love with their creations, and system designers do too. Evolutionary computing can help engineers overcome their design infatuations.

Born in 1963, Qualls grew up in Memphis, Tennessee. His mother came from a farm in rural western Tennessee. Qualls's maternal grandfather was a creative jack-of-all-trades—a welder, mechanic, and machinist all in one. "A lot of what I do today is similar," notes Qualls. "I figure out how things work—take it apart and hope I can get it back together." Qualls's father was a salesman and instilled a strong work ethic in his son. "He said 90% of success is getting to work on time, dressing decently, and working hard," remembers Qualls. Mindful of his father's admonition, Qualls often expresses his appreciation of his good fortune to end up at Oak Ridge.

Qualls attended public school in Memphis and then Harding Academy, a Christian private school. He describes himself as a "middle of the pack" student. Qualls became interested in industrial arts, which encompasses architecture, engineering, and physics. In 1981, Qualls began the architecture program at the University of Tennessee in Knoxville. He soon discovered that he couldn't draw very well and became more interested in a building's structure than in what it looked like. He considered an engineering degree. The question was what kind—mechanical, electrical, civil, or aerospace; there were many possibilities. Qualls decided to interview the head of each department. His talk with the chairman of the nuclear engineering department, Pete Pasqua, convinced him to choose nuclear engineering. "He sold me on the idea that nuclear engineering was important, something significant to do with your life," recalls Qualls. Nuclear engineering had the added benefit that fewer people would be in the field. The other engineering specialties had two or three hundred graduates per year, while the nuclear group had about twenty. Qualls felt that if he went to Pasqua's department, he would have at least one person who cared about his success.

Nuclear engineering appealed to the tinkerer in Qualls. It involves math, computing, engineering, thermal hydraulics, machines, pumps, and turbines. His solution to a simple puzzle in a first-year computing class would be a harbinger of his current design techniques. The class assignment was to write a program to solve the peg puzzle often found on tables in Cracker Barrel restaurants in the South. The board starts with 15 holes and 14 pegs, so one hole is left empty. Players proceed by jumping over pegs into empty holes, removing the jumped peg each time. Solving the puzzle means ending with just one peg in the original empty hole.

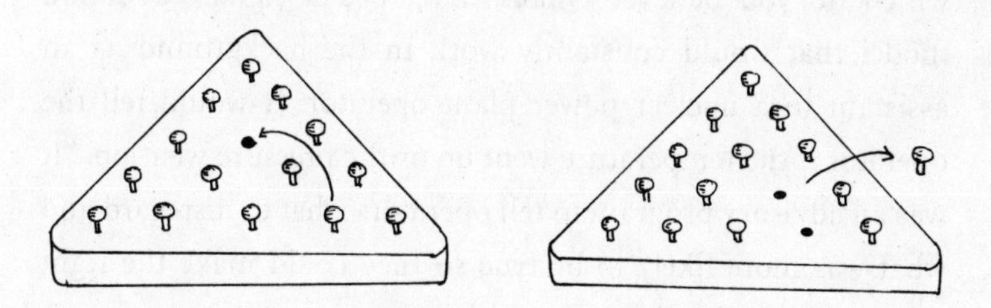

Each jump in the peg puzzle causes a peg to be removed.

For the assignment, Qualls wrote a program in Prolog, a language that embeds backtracking (see the box "Keep Trying"). "It was the neatest thing. The computer wasn't real bright, but it never gave up. And you could set it up to keep trying things in a systematic way so that it would eventually find a solution," remembers

Qualls. He admits the solution wasn't elegant, but some things can't be solved elegantly. It was his introduction to the fact that computers could be used to grind out a solution if you posed the right question.

In 1984, five years after the Three Mile Island accident, Qualls began work on his master's thesis on signal validation in nuclear reactors—determining how different sensors communicate with one another and which ones to trust. In the Three Mile Island accident, the core of a reactor had partially melted. Only the integrity of the reactor vessel and containment structure had averted a widespread release of radiation. Expert analysis of the accident suggested that the operators had made poor decisions because they were overwhelmed with sensor readings, some of which were incorrect.

If you have a control panel with two conflicting sensors, which do you believe? Qualls developed a signal validation model that would constantly work in the background as an assistant to a nuclear power plant operator. It would tell the operator if the temperature went up or the pressure went up. "It was an advisory program to tell operators what to disregard and what was more likely to be true so they could make the right decision," says Qualls.

To inform the advisory program, Qualls used various engineering rules of thumb. For example, sensors made by the same manufacturer and calibrated by the same engineer should give similar readings. If they don't, then the outlier—the reading that is outside the norm—is probably wrong. History can also help. If the pressure in vessel A is always higher than the pressure in vessel B when a certain pump is on, but the pressure sensors indicate otherwise, then something may be wrong with the pump.

Keep Trying: Backtracking 101

A technique that could work for the peg puzzle, Sudoku, or any similar game is called ***backtracking***. In the peg puzzle, backtracking entails the following idea: The board is in a state *s*. You choose a peg that can jump and one of its jumping target positions, and you make the jump. The result is a new state, *s'*. You keep making choices until you reach a state where you can't jump anymore. If it's the winning state, then you're done. If not, then you go back to the last state in which you had a choice, and take a jump that is different from any jump you took previously.

Backtracking is the process of going back after a failure, recovering the state, and trying something new. To do this systematically, your program must remember every previously visited state, as well as every jump the program has already tried from that state. In Prolog, the language Qualls used, a programmer can state a goal and the compiler will direct the hardware to find a solution by backtracking.

In the Three Mile Island accident, a valve that was supposed to close failed to do so. Qualls's program might have helped operators figure this out.

Qualls believes that human operators still have their place, but computational support can help. Unlike the human operator, a computational monitor never loses focus and never falls asleep after a night of partying. "It's always looking for something, and that's all it does. It can also be programmed to learn—to continually learn [by building up databases of events] what's normal and what's abnormal," says Qualls.

At the time of Three Mile Island, the nuclear power industry did not use advisory programs, but they do now. Many fields have come to discover the same delicate computer-human balance. Computers monitor mostly unchanging information and alert operators about changes. Humans synthesize the results in a broader context and make decisions. When too much information arrives at once in time-critical settings, humans need help. Organizations often use checklists and expert systems to help people troubleshoot. At a nuclear power plant, checklists guide operators to preserve the integrity of the reactor core whenever problems occur.

After completing his master's in nuclear engineering at the University of Tennessee in 1988, Qualls moved to Oak Ridge National Laboratory to do his doctoral thesis in 1991, and he's been there ever since. For his thesis, Qualls worked on a fusion device called a *stellarator.* (Fusion is the energy source of the stars; *stella* is Latin for "star.") Qualls's job was to build the pellet injector, the device that shoots fuel into the doughnut-shaped plasma confined by the stellarator.

No school ever prepares a student completely for a big engineering project in which so many components and people have to work together. The first project helps a young design engineer appreciate what it takes to make such projects succeed. Qualls still marvels at how helpful the Oak Ridge staff were to an aspiring scientist: "They're the smartest people you would ever want to work with. And none of them would tell you they're smart. They don't think like that. They're just bright people who come to work every day and do what they have to do. If you listen to them, you just learn so much."

The work at Oak Ridge is far-reaching. Vestiges of its atomic bomb history are still evident in the continued cleanup of radioactive wastes and the ongoing endeavors of the adjacent Y-12 National Security Complex, which works on nuclear weapon components. Since being taken under the wing of the Department of Energy in the 1970s, the staff of 4,300 at Oak Ridge have been tackling "Grand Challenge" problems, ranging from neutron science to the creation of nanophase materials. Spread over 33,750 acres, the Oak Ridge facilities provide a venue for experimentation and the exchange of ideas among the world's foremost scientists.

Qualls fits in well with the wide-ranging scope of the place. He has acquired a reputation of being able to pull different components together to deliver a result that meets a variety of difficult requirements. For example, in its forward-looking work, NASA plans for nuclear-powered habitats on the moon. Qualls has to consider fuel, the nuclear core, the containment vessel, the protective shield, power conversion equipment, and electronics. He's not an expert in any of those areas, but when he works on a project he consults with experts in each one. He brings everyone together in a conference room and elicits opinions. As Qualls describes it, "By the time I get all the way around the room, I've got my system. I put it together. If it doesn't work, I start changing things."

Like many other people, Qualls starts with an initial concept and then proceeds to tweaking. But how do you choose a good initial concept? And once you have that concept, how do you tweak it? Qualls uses a genetic algorithm approach starting with hundreds of random initial designs. Tweaking involves searching for improvements; for instance, can the fuel system be made more efficient, perhaps at a slightly higher cost in volume? Sometimes no tweak will improve overall performance, but new possibilities for improvement open up. For example, a different pump material might weigh the same as the original pump material but allow greater flow and higher energy efficiency. The ideal tweaking strategy involves searching for improvements but also looking at neutral or even negative changes to see if they open up new possibilities for later improvement. Besides tweaking each design individually, genetic algorithms allow different designs to combine with one another.

The genetic algorithms/evolutionary computing strategy, when applied to design, works as follows: (1) start with many initial designs, (2) tweak individual designs with random changes, and (3) combine pairs of individual good designs to get better ones.

Before he began using genetic algorithms, Qualls would take three weeks to come up with a preliminary design. "And I made everybody real mad at me because nobody got what they wanted," Qualls admits. By contrast, he says, "on the computer all I have to do is change my constraint, push the button and go away." Genetic algorithms share this "push the button and go away" benefit with other computerized design methods, such as backtracking (see the box "Keep Trying" on page 47). As long as the designer encodes the constraints properly in the computer model, adding a new constraint requires nothing more than rerunning the model. Unlike a human designer, the genetic algorithm model will not try to minimize changes in a design to which it has become emotionally attached. It can start with a completely new set of initial designs and perhaps arrive at something completely different.

If you doubt that humans preserve their biases as constraints

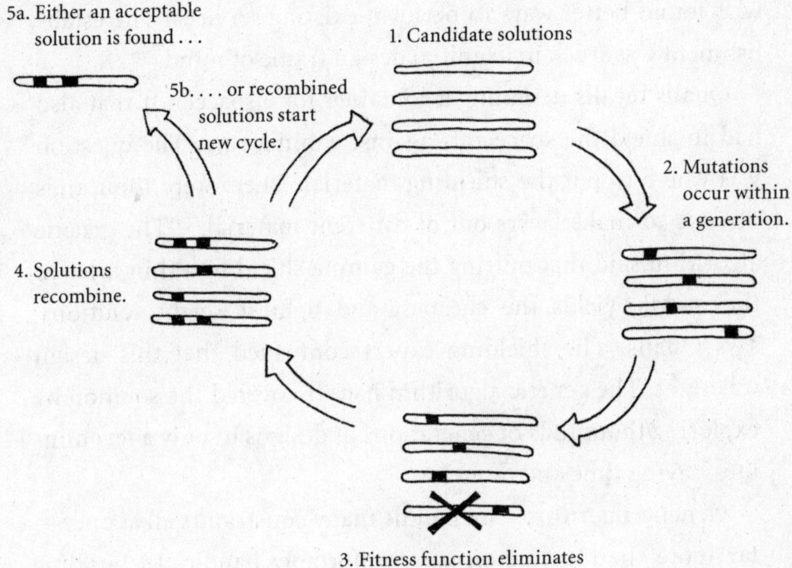

The basic cycle of genetic algorithms. (1) Start with a set of candidate designs, each one represented here as a chromosome consisting of a sequence of genes. Each "gene" is a particular design choice (for example, the shape of the reactor container). (2) The genetic algorithm changes a few genes in each chromosome through random mutation. (3) The algorithm evaluates the chromosomes, eliminating low-scoring ones. (4) The remaining chromosomes are recombined (though the fittest one is retained in case nothing better is found). (5a) The cycle stops if a good enough chromosome is found. (5b) Otherwise, the cycle starts over.

when technology changes, just remember how many years passed in the early 1900s before cars no longer looked like "horseless carriages." Consider also that many recent business and technological innovations—express mail, microprocessors, the Web, and cell phones—arose from people outside of the establishment who found better ways to perform existing services. The establishment was stuck in its initial design frame of mind.

Qualls recalls designing a container for a spacecraft that also had to shield the spacecraft against gamma rays. The question was where to put the shielding material. There were thousands of ways to make layers out of different materials. "The genetic algorithm said that putting the gamma shield in a thin layer in the middle yields the cheapest and lightest-weight solution," says Qualls. The shielding expert confirmed that this design was right. The genetic algorithm had discovered the solution by exploring thousands of generations of designs in only a few minutes, saving time and money.

Genetic algorithms can handle many constraints all at once—far more than a human can comfortably handle. In building nuclear power plants, for example, designers have to deal with the problem of radioactive waste. Disposing of nuclear waste is a complicated issue involving many processes, capital investment, manpower, and risks. "The genetic algorithm is going to let us take those larger, more detailed models and do an optimization on them and find good solutions within that larger design space," predicts Qualls.

Genetic algorithms also enhance the teamwork on a project. In the past, when Qualls decided on the compromises for a design point, many specialists disliked the result. But when a genetic algorithm churns out 100 designs, the specialists begin to see a pattern of compromises. People's attitudes improve. "The algorithm helps different subsystem experts understand the constraints of others," says Qualls. As a result, there's more cooperation among team members.

Despite his reliance on genetic algorithms, Qualls remains a firm believer in the importance of human designers. They can verify that the computer-generated designs work. In effect, the designer presents the computer with a landscape of possibilities and allows the computer to optimize among them. So, people still have a starring role. "A genetic algorithm can optimize an airplane if it has an airplane model," Qualls says. "But it can't choose between an airplane and a helicopter unless it is given a model of both."

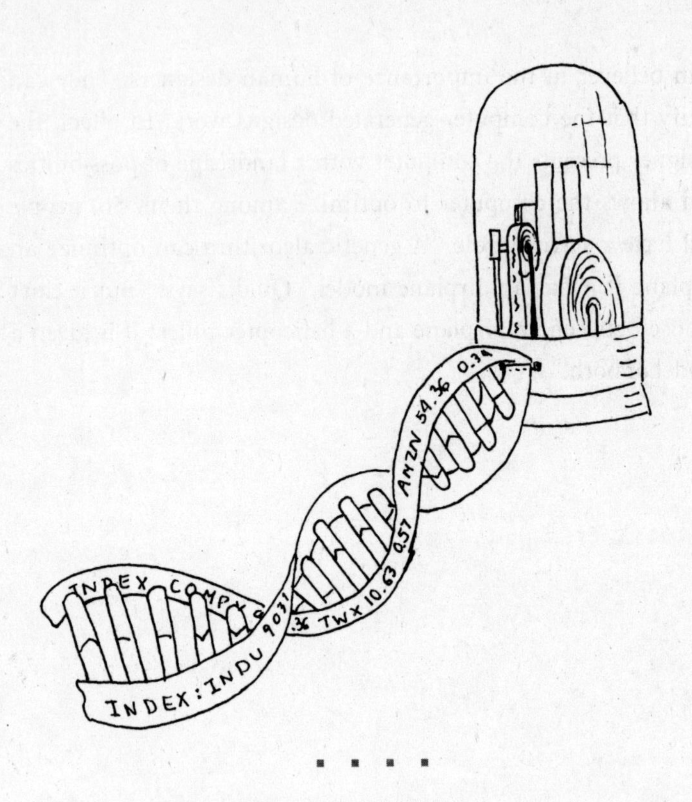

■ ■ ■ ■

The market goes up and then goes back down—for no apparent reason. You see it all the time. It seems repeatable, at least partially. The finer the time series, the more repeatable it appears. Next year, Microsoft's price might be double or half of its current value, but in the next microsecond it's only going up a penny or down a penny.

— *Jake Loveless*

Chapter 4

JAKE LOVELESS and AMRUT BHARAMBE

■ ■ ■ ■

Riding the Big One

A FEW YEARS AGO, A POSTER ABOUT A HAND-BUILT–AIRPLANE festival displayed this tagline: "It takes brains to build one and a completely different part of the body to fly it." So it is with financial markets trading. You look at history and try to find patterns. You prepare to act on those patterns, but you never really know what is going to happen. Then you have to take the plunge and commit to a strategy. Your brain is overridden by guts, instinct, something else. What if you bet $10 million on a pattern that doesn't work? Even if it's not your money, your job and reputation hang on the outcome.

Program trading, using computers to trade large blocks of stock, has been around since the 1980s. What's newer is the study of shorter-term trading, known as *micromarket trading,* and the need for adaptive rules that can act without human intervention. Evolutionary computing finds new rules and changes those rules as conditions change. It took a surfer dude and a former tennis star to figure all this out.

Born in 1980 near Miami, Jake Loveless grew up in Boca Raton, Florida, and was raised by his mother and stepfather, both teachers. His mother specialized in math. To test Jake's mental agility, she had the "bad habit," he recalls, of explaining quadratic equations on a piece of paper, ripping it up, and then asking Jake to replicate what she had done. "You had to know them inside out or derive them yourself," he remembers.

Jake enjoyed chemistry and math in high school, but he decided to study philosophy at the University of Florida. He wasn't sure what he was going to do with his life, but one activity caught his interest: surfing. He even toyed with the idea of becoming a professional. He audited a lot of math courses and found calculus boring, but he loved chaos theory. His academic performance was short of stellar. "I failed statistics and C programming—twice. It was the only programming course I took. So my second year, I switched to scuba diving," Loveless says.

After bombing out at the University of Florida, Loveless thought he would try a change of venue. He moved to Virginia to be with his father and went to work in the warehouse of a small moving company in Fredericksburg. His job consisted of taking six pieces of plywood and nailing them together—building boxes instead of quadratic equations. He was soon promoted to furniture moving.

Loveless became interested in how the moving business worked, and he wanted to figure out the best way to make money. He realized that with a low-margin business like trucking, profits depended on scheduling. He wrote up a business plan and borrowed a suit. When the CEO of the company visited, he arranged a meeting and went to see her. Impressed by Loveless's scheduling plan, the CEO made him the first director of technology and put him on salary, a big step up from the $4.25 an hour he had been earning.

Loveless's goal was to keep the trucks full. The software, written in Cobol, one of the oldest programming languages, exploited the fact that if a truck went from Virginia to California, it would make more money dropping off and picking up along the way than going to Tennessee empty and continuing to California. It was a classic case of optimization. Loveless had realized that using fewer trucks with more efficient routing has a tremendous impact on the bottom line, especially in an industry that in the year 2000 was still technologically backward. "You're talking about an industry where there weren't a whole lot of mathematicians," says Loveless.

Loveless joined the American Moving & Storage Association and continued to develop cross-company shipping software. Then he got a call from Appian Corporation, a technology company in nearby Vienna, Virginia. Appian designed database systems for the US Navy—an enormous undertaking. The company's comparative advantage was speed and scalability. Appian's database program was based on a programming system known as K (see the box "APL, K, and a Manifesto about Computer Science Education"). Though he had failed his class in the much simpler C language, Loveless took immediately to the power of K, expressing the mantra of so many K programmers: "I don't like to write a lot of code."

In 2003, after working at Appian for two years, Loveless and fellow employee Mark Sapnar started their own company. Working out of a garage, they continued to consult for the Department of Defense. When the company was sold, Loveless found himself back in Florida surfing (the wet, not the Web, kind) and consulting for companies such as Microsoft and Red Hat.

APL, K, and a Manifesto about Computer Science Education

In 1955, a pioneer in computing, Professor Howard Aiken, convinced the Harvard administration to introduce a master's program in "automatic data processing"—basically, computer science. Aiken hired Ken Iverson, a young Canadian graduate student, to teach some of the classes. Iverson found that the mathematical notation of the day was imprecise, so he began to develop his own notation. He borrowed vector and matrix operators from linear algebra, and he encouraged programmers to think about collections of information instead of single information units. In 1979 Iverson won the Turing Award, the Nobel Prize of computer science, and many language designers have expressed their debt to his language, APL.

Despite these accolades, APL never hit the mainstream, and it has been mostly ignored by universities. First, APL literally looked like Greek—it's full of Greek characters. Writing in APL requires a special keyboard or a remapped keyboard and a great human memory. Second, the operators are so powerful that they frightened conventional programmers. APL can manipulate million-by-million matrices with only a few keystrokes, but conventional programmers are used to manipulating one number at a time.

In the early 1990s, the K language, based on APL, and another language called J, was designed and implemented by Arthur Whitney. It uses normal keyboards, so the symbols look less scary. The keys have abnormal meanings; for example, + means "plus," but +/ means "sum over a whole list." Whitney made K extremely efficient by using clever data layouts, by implementing a variety of efficient sorting algorithms (some for floating point numbers and others for small domains of values), and by keeping the implementation small.

Whitney extended the power of APL's operators substantially. As a result, K can handle everything from communication to database functionality to linear algebra. Applications that normally require the integration of Perl, Java, and SQL can be written in a single language. (Coauthor Shasha uses K for all his work. We are admittedly biased.)

As of this writing, languages based on Pascal, C (C++ and Java), and sometimes LISP dominate the university computer science scene. Data-rich disciplines, such as financial mathematics and biology, teach vector languages such as Matlab and R.

Computer science teaching must change for the simple reason that multicore chips demand a change. ***Multicore chips*** are printed with many processing elements, each of which is capable of running a program on its own. Writing a sequential program for a multicore chip implies that only one core will do any work. Languages like Java enable programmers to write multithreaded programs that can use all cores, but most programmers do this poorly, producing buggy applications that could easily lead to loss of life or property. It makes much more sense to program in a language whose operators deal with matrices and database tables as single units and compile it to multicore. So here's our manifesto: we need to teach super-powerful languages to students.

In the course of his consulting work, Loveless met a naval officer named David Jacobsen at the Navy Medical Information Center. Jacobsen's job was to analyze root cause epidemiology, to link medical side effects with possible causes. He was using an algorithm called DAMI, as in "DataMiner." It was Loveless's first exposure to a genetic algorithm. As it churned data, it used a very childish graphical interface. "But behind it was an unbelievable search algorithm," remembers Loveless. "I had never seen anything crunch data like it did."

Jacobsen was pitching the algorithm to advertising firms and for medical applications. Loveless suggested another market. He contacted a friend from Appian who was working at Cantor Fitzgerald, the Wall Street bond-trading company sadly made famous because so many of its employees died in the attack on the World Trade Center on September 11, 2001. In contrast to the popular image of investment bankers in power suits, Wall Street trading organizations are casual. Every trader faces the same test: make money for the company and we'll reward you; fail to make money and we'll fire you. Looks don't matter.

When Loveless and Jacobsen showed up, the company handed them a huge pile of US Treasury data and asked them to find profitable patterns. Loveless admits, "We had no idea what US Treasury bonds were, other than our grandmother might have given us treasury bonds when we were kids. Bids, offers—we didn't know what any of this was." To fill in the gap in their knowledge, Loveless got to know a Cantor Fitzgerald employee named Amrut Bharambe.

Bharambe grew up in Nagpur, a small town in central India. Starting at age seven, he played professional tennis six to nine hours a day, traveling all over India. But at age fourteen, he incurred debilitating knee and back injuries. Forced into premature retirement, Bharambe applied his competitive nature to school. His father, a statistician, and his mother, a PhD in education, helped him catch up. He completed his studies in mechanical engineering in India in 2001. He wasn't sure what he was going to do with the degree. "Every Indian would go to the US and the UK and make a lot of money from this whole math finance thing," Bharambe relates. Following that lead, Bharambe came to the United States and ended up at New York University's Courant Institute of Mathematical Sciences to study for a master's in mathematical finance. An internship led him to Cantor Fitzgerald's trading desk, just a few months before Loveless arrived.

At the time, Cantor Fitzgerald ran the exchange for US Treasury bonds, so in principle it was not taking any risk, but that was about to change. The firm created a Debt Capital Markets (DCM) division, the first division at Cantor to take partner capital and place it at risk. Within DCM there were eight traders and several managers. Bharambe was hired as a consultant. For the previous decade, Wall Street had been program-trading equities. But nobody had looked into doing the same thing for treasuries. DCM offered the possibility of a new instrument and a new exchange. "David Jacobsen and I mined the data and found all these interesting patterns," says Bharambe.

Meanwhile, Loveless continued working as a consultant. "I was sitting on the surfboard in Florida, and I had rigged up a phone. I took consulting calls out at the beach. My wife called one afternoon and said 'Are you on the beach?' I said, 'No, I'm in the ocean.' She said, 'You need to go back to work.'" Loveless called the CEO of Cantor Fitzgerald and asked if he could work there as a full-time employee. Loveless and Bharambe were then asked to set up Cantor Labs and expand program trading. Faced with the massive quantity of data, Loveless decided to use K. He really didn't know anyone in New York, but he found a welcoming community of K programmers, "among the best and brightest in computer science. The entire back end of Millennium, one of the largest hedge funds, was built by a single K programmer." Loveless thought he had the right tool.

Bharambe and Loveless began to build an entire trading system for treasuries—from tracking the market price stream to sending in the orders. The key remaining part was to find good rules. All data-mining applications involve the same strategy, whether on Wall Street or elsewhere: Take historical data and divide it into two parts—the training set and the testing set. Then find rules in the training set and see whether the rules hold in the testing set. The data is divided in this way to avoid formulating a rule that arose just by chance. Even random pieces of data appear to have patterns. The test data is there to determine whether the rule is likely to be real.

A typical rule might be this: "If the price of a 10-year treasury rises by a certain amount over 2 minutes while the 5-year treasury doesn't move, then the 5-year treasury is likely to rise in the next 2 minutes." Rules that survive the test data are said to be *backtested.* Statistics can also help evaluate when rules hold by pure chance (bad) or are likely to be repeatable (good).

Because they were short-term traders, Loveless and Bharambe wanted to find rules that, using the activity in the previous few minutes as a guide, could predict the market's moves in the next 5 minutes or less. But they weren't sure what kind of rules to look for. Rules might be made up of *attributes* such as the slopes of the market price line over 1- or 5-minute intervals. Or rules could be constructed out of moving averages over various time periods. Each attribute has multiple values. A *trading rule* is a collection of these values.

Since Loveless and Bharambe didn't know which attributes would be valuable—there are many ways to describe a time series—they decided to start with the simple concept of slope. They then had to choose which slope, market data fields, and time periods to combine into rules. "Amrut just handed me 28,000 attributes—the slope of the traded price over 2 minutes, the slope of the bid volume over 30 minutes, and so on. He said, 'The answer's in there somewhere,'" recalls Loveless.

The "brute force" approach to finding rules consists of discovering all combinations of attributes and values for a desired outcome, such as maximum historical return. A good rule would find conditions that lead to a price rise in the training set but not to a price decline in the testing data. A rule that contained thousands of attributes would be useless, however, because it might never or rarely recur and would not survive the statistical tests. The computational solution is to find rules that use relatively few attributes and maximize the desired outcome. The trouble is that even a small set of 10 attributes, in which each attribute has 10 distinct values, yields over 10 billion combinations. To solve this dilemma, Loveless explains, "We needed to think of something that was autonomous and could churn through billions of combinations. We didn't have time to think of what the actual relationships were."

Loveless had read enough to know what an evolutionary search system would do. The system would start by randomly selecting attributes and values, build a trading rule from them, backtest them against the training set, and record the profit or loss. The next step was the difficult part: how should those results guide the next attribute and value selection? For example, suppose profitable rule 1 stated,

> If the 1-minute slope of the traded price is >5%, and the 5-minute slope of the traded size is >10%, then buy.

and profitable rule 2 stated,

> If the 2-minute slope of the traded price is >5% and the 10-minute slope of the traded size is <20%, then buy.

What would a new rule built from the two look like? And how would you combine them? You could randomly interchange components of each of the two and end up with rule 3:

> If the 2-minute slope of the traded price is >5%, and the 5-minute slope of the traded size is >10%, then buy.

Or you could modify a profitable rule by shifting one of its values—for example, changing the value of the 2-minute slope attribute in rule 2 from 5% to 10%, resulting in rule 4:

> If the 2-minute slope of the traded price is >10% and the 10-minute slope of the traded size is <20%, then buy.

Loveless used a collection of rule-searching strategies with technical names like *classical elite selection*, *local shifting*, *tabu-constrained crossover*, and a few he invented along the way. He combined these with different fitness criteria. Each method attempts to build a collection of solutions. At the end of each *generation*, or iteration, all solutions from all methods are evaluated and recombined.

Some of the recombination methods included simply switching out one attribute and value for another with a higher or lower score, but Loveless decided he wanted a more general system. "In the end we built a generic evolutionary computing framework where many recombination methods and different optimization functions execute in parallel," he explains. At the end of a generation, all of the search methods share their results, and a second-level heuristic then allocates resources to each search method for the next generation.

As of this writing, Loveless and Bharambe have about 15 distinct recombination methods and dozens of optimization functions to which they add fairly often, with interesting results over the years. For example, they found that a search type whose optimization function was to look for the lowest profit actually provides useful attributes and values for assimilation into other search methods that are looking for the highest profit. "Evolutionary computing takes you to strange places," comments Loveless.

To be useful in a variety of situations, the genetic algorithm must find thousands of promising rules. Each generation tests over a billion rules per minute on each processor. The genetic method gives no guarantees about finding the most profitable rules, but it popped out a lot of them. Unfortunately, Loveless and Bharambe couldn't understand most of them. "At 10, 20, or 30 attributes, you just can't keep it in your head. We realized that in the end, this would be a truly black-box system. We wouldn't know if it was going to buy or sell," admits Loveless.

Finally, they went live. They turned the algorithm on at 9:00 AM on a Monday and ran it for a short time. The system bought and sold $10 million worth of securities. "I was sweating. I used to bring two shirts to work for the first couple of months. It was exhilarating in a jumping-out-of-a-plane sort of way," says Loveless. And it worked.

The genetic algorithm works well with all sorts of large data problems, and it still did throughout the financial turmoil of 2008. But the surfer dude inside Loveless yearns for new crests. He next wants to tackle computational biology, such as mapping the interaction between proteins (for more on this, see chapter 12, on David Shaw). "That's the only thing that has more data than Wall Street."

■ ■ ■ ■

Most engineers are taught to look at things from the viewpoint of what you want a system to do. What about the things you don't want it to do? That's all that safety is. Forcing people to think about what they don't want the system to do.

— *Nancy Leveson*

Chapter 5

NANCY LEVESON

. . . .

"It's the System, Stupid"

ON APRIL 10, 1912, THE *TITANIC* LEFT ITS PORT IN SOUTHAMPTON, England. The newspapers celebrated a masterpiece of ship design—three propellers, four smoke towers, radios, electric lights, and a swimming pool. Some of the world's wealthiest industrialists were on board, sequestered in their ornate first-class cabins. Four days later, the boat collided with an iceberg and sank, with great loss of life. How could such a "perfect" vessel succumb to this ignominious fate?

A postmortem revealed that the boat had been traveling fast at night in the iceberg-laden North Atlantic waters, yet the watch for icebergs had been intermittent. In addition, the boat was carrying enough lifeboats for only about half the passengers. Worse yet, when the 40-person lifeboats launched, some carried only 12 passengers. In the aftermath of the tragedy, some people blamed the captain and the lookout. Others believed that the constellation of failures in design and procedure had been just astronomically unlucky. Another viewpoint held that the ship had sunk as a result of overconfidence—from the belief that safety precautions (sufficient lifeboats, proper lookouts) were unnecessary on a "practically unsinkable" ship.

Though technology continually improves, accidents will always be with us. In fact, more advanced technology means enhanced destructive potential. Compare the consequences of a meltdown at a nuclear power plant with a blade slipping off a windmill. Nancy Leveson of MIT's Department of Aeronautics and Astronautics analyzes accidents, especially life-taking ones. She believes that looking for proximate causes—reckless captains or inattentive lookouts—makes future accidents *more* likely. She maintains that the focus on proximate causes can often obscure the systemic ones.

Leveson has an unusually broad background that helps her look beyond technological factors. She considers the system writ large—technology plus management, industry and company culture, government, society, and economics—what she calls "socio-technical" factors that affect safety. She includes the entire life cycle of systems, from concept development to design, deployment, and finally retirement. Many holistic approaches to safety drown in a sea of words, but Leveson's methodology offers the possibility of improved safety practices for complex systems.

Born in 1944, Leveson grew up in West Los Angeles. Her father, the youngest of ten children, was the only one in his family to go to college. After working for an accounting firm, he retired early and stayed home to manage his investments. Leveson's mother ran her own sportswear company. Shy and bookish, Leveson had taught herself to read (at age three); and she loved exploring the encyclopedia, which was bigger than she was. But

Leveson's mother wanted her to be more social. "I had to hide because my mother would push me to go play with the neighborhood kids," Leveson remembers.

After excelling at an IQ test in the second grade, Leveson skipped a year in school. That made her younger than everyone else in the class—not an ideal situation for a shy kid. To make matters worse, Leveson was bored. Like many other bright children faced with uninspired teachers and an unchallenging curriculum, Leveson sought other intellectual outlets. "Most of the time I spent the hours daydreaming and creating alternative worlds," she recalls. After daydreaming her way through high school, Leveson went to UCLA in 1961. She studied math because that major required so few courses. She could then take other courses in many different subjects, including the social sciences, humanities, and arts. She credits her approach to research to this very broad education.

At UCLA, Leveson became more social. It was the mid-1960s, a time of great political and social upheaval. Students gravitated toward activities that connected them with the world beyond academia. Leveson skipped most of her classes during her last two years at UCLA and studied on her own. She just showed up for the finals.

Most universities offer a mixture of theoretical and applied mathematics; there is often a tension between the two disciplines. UCLA's math department was proudly theoretical, but it did offer a one-hour-per-week course on computer programming, on an obsolete IBM 1620. Following her modus operandi in her other courses, Leveson signed up for the course and then forgot she was enrolled. A week before the finals, she got a notice telling her the time and location of the exam. She asked the professor to let her drop the course. He was sympathetic but denied her request, and he suggested that since the final was open book, perhaps she could pass it anyway. Leveson reasoned that it didn't make sense to do poorly in her other classes by trying to learn about computers in a week. She went into the final exam stone-cold and started reading the textbook for the very first time during the exam. She failed the course. "I have the distinction of being perhaps the only PhD in computer science who failed her first computer class," admits Leveson. While sitting at the exam for three hours and reading the textbook, she decided that computers might be interesting. She decided to pursue computer science as a career. "It was like solving puzzles, which I loved," says Leveson.

Her next problem was how to talk her way into graduate school, given her less-than-sterling academic record. There was no computer science department at UCLA at the time, but there was a program with less stringent requirements in UCLA's School of Management.

She approached a professor who did research in artificial intelligence—Earl "Buzz" Hunt. He agreed to take her on if she did well in his graduate operating-system and AI classes. Unlike the other students in the PhD program, Leveson knew nothing about computers. She was also the only female in a school of eight hundred. "I was scared," admits Leveson. "I had no idea what was going on in my classes." For once, she read the course books and asked questions. "I had my hand up the whole time," she recalls. "Everyone in the class thought I was going to fail. I thought I was going to fail. The professor thought I was going to fail. This 'girl' is totally in the wrong place." Instead, Leveson proved everyone wrong and got the highest grades, and Hunt agreed to supervise her PhD. But Leveson felt uncomfortable being in a place with so few women. She had never had a female professor, and remaining in academia seemed inappropriate.

After completing her master's degree in 1967, Leveson decided to work for IBM in Los Angeles as a system engineer. At the time, IBM offered one-stop shopping to its customers, providing hardware, programming, and even operations support. Leveson did all of that, plus debugging operating systems and teaching others about new tools in information and database systems. Leveson enjoyed her work and was doing well financially, but she grew restless. With money in her pocket, she quit IBM and decided to trek around the world.

For two years, Leveson hitchhiked her way one and a half times around the planet and lived on an average of 50 cents a day. While she was staying with an Australian couple in the highlands of New Guinea among Stone Age aboriginal tribes, she began to take stock. She contrasted her life in the United States with what she saw in Asia. "In the West we feel that the world revolves around us as individuals. We think our own lives are very important. When you're in Asia and there are all these masses of humanity living under terrible conditions, you start to feel very small and unimportant in the larger scheme of things," Leveson says. She felt it was time for her to do something that would make a difference and give meaning to her life.

Leveson returned to the United States to teach high school math. After a year, she decided she wanted to work with emotionally disturbed kids. She enrolled in a program in developmental psychology at UCLA. Although she did well on the academic side of things, she found working with the kids depressing. Another negative was that it would take her eight years to get a degree in cognitive psychology, a field that offered few jobs. Some friends suggested that she go back to computer science and get a doctoral degree in three years. She switched back to computer science, in artificial intelligence, but she fell out of love with it. She felt that 1970s AI researchers were stuck with the same issues they had faced in the 1960s.

Leveson switched once again, this time to software engineering and programming languages. She finished her degree in 1980 and accepted a job at the University of California at Irvine. She was the first woman in the department. "It wasn't easy to be a 'first,'" Leveson says. She also couldn't publish her dissertation, because it had been plagiarized by someone who had asked for an early copy. "I was pretty bummed out," recalls Leveson. During her first week at Irvine, Leveson got a call from an engineer at Hughes Aircraft. The staff at Hughes were working on a torpedo with 15 microprocessors. They had some concerns—not about missing the enemy, but about the possibility that the torpedo would turn around 180 degrees and hit the launching vessel, the ultimate insult in friendly fire. Leveson told the Hughes engineer that she worked on the formal semantics of programming languages, not reliability. "I don't know anything about torpedoes; I'm an applied mathematician," she told him. "He said, 'We've got some money.' I said, 'I'll learn,'" And so Leveson's education in system safety began.

In commercial aviation and nuclear power, safety engineering focuses on making systems "fail-safe." Mainly, this means making each component very reliable and then having several identical (redundant) copies of the components to make them even more reliable. Engineers in commercial aviation and nuclear power also emphasize learning from past accidents—what could be called the "fly-fix-fly" approach. This approach works well because the system designs are relatively simple and change very slowly. Until quite recently, computers were used very conservatively, if at all.

The safety problem in the defense industry is very different. Weapons are always being updated with the latest technology. Since the early intercontinental ballistic missile (ICBM) system of the 1950s, computers had been used to control very complex systems. Accidents and near misses were frequent, even though components were built well and there was redundancy everywhere. Engineers discovered that in complex systems, many safety problems arise from the interactions among components, not from the failure of individual components. Redundancy isn't enough—it doesn't provide protection against software errors. Multiple versions of incorrect software do not result in correct behavior.

To cope with safety problems in complex defense systems, a new form of safety engineering was developed, called *system safety*. System safety builds safety into the design from the beginning rather than adding it on after the system is built. System safety focuses on *hazards*, or unsafe states of the system, such as an inadvertent missile launch. Hazard analysis determines how such dangerous states arise. It is essentially investigating an accident before it happens. The design is then changed to eliminate the potential causes of the hazards.

Leveson's diverse background, which included the social sciences, became a plus. She was able to integrate engineering with the social and cultural aspects of accidents, and her computer background helped her tackle complex system-engineering problems. "The digital revolution was taking engineers by surprise and changing the bedrock under their feet," says Leveson. Many of their basic assumptions, such as that different components fail independently, would not hold for computers and software, where interconnected system failures occur all the time.

Like most other safety analysts, Leveson tried to extend traditional safety-engineering techniques to software. She thought about testing software for safety, but she found that approach doesn't work. "Accidents usually occur because of things you forgot or because the assumptions about the system and its environment are incorrect," notes Leveson. The testing process itself can be plagued by erroneous assumptions, and it comes too late. Making significant changes to complex software at the testing stage vastly increases delays and costs.

Leveson turned her attention away from software testing to software design. But then she realized that most software-related accidents stem from errors in the requirements, not in the software design. This should come as no surprise: the *Titanic* had incorrect requirements—if the designers included lifeboats, why not enough for all the passengers and crew? So Leveson tackled the requirements field and also looked at human-computer interaction because operators were often blamed for accidents that really resulted from the software having confused or misled them. "Once I got into human factors and into requirements, I realized I had backed my way out of computer science and into system engineering," explains Leveson.

Because computers allow enormously complex systems to be built, they have created new problems in system engineering. "We no longer have physical laws that limit the complexity of our designs," says Leveson. Before software became pervasive, engineers used physical constraints to ensure safety. For example, a classic train design has the brakes engage by falling on the wheels in the absence of electric current. Gravity provides safety when there is an electrical failure. No physical force ensures safety for software-guided torpedoes, air traffic control systems, or financial regulation. The versatility of software discourages fail-safe design.

In her "ever-expanding search for worse weather," Leveson left Irvine for the University of Washington in Seattle and then MIT's Department of Aeronautics and Astronautics. She fell in love with MIT because it was the first place where she found a group of people thinking about systems in a larger context. Leveson is now working on what she calls the socio-technical aspects of large systems, which integrate management and social sciences with engineering. Her newest book is called *System Safety Engineering: Back to the Future.* She wants to bring the ideas of the early system safety pioneers back into the forefront of safety engineering to achieve a new way forward.

Leveson faces an uphill battle: "Engineering students are taught about reliability, but not safety," she says. There is no formal education in system safety. Engineers have to learn on the job from other system safety engineers. Reliability and safety may sound similar, but they differ greatly. For example, if you examine a valve in a nuclear power plant, you can determine that the valve is reliable but you can't predict whether the plant will be safe. "Safety depends on interaction. In system theory, safety is considered an emergent property," Leveson says.

An ant colony is an often-used example of the concept of emergent properties. The colony can survive even though individual ants die often. The colony members know how to carry out specific responsibilities, such as gathering and stockpiling food. Similarly, in computer systems it is possible to achieve safety using unreliable components, provided there are enough of them and they interact properly. By contrast, in non-emergent systems even a minor mistake can lead to a terrible accident.

Engineers traditionally ascribe accidents to a *root cause* in the chain of events leading to the accident. Leveson argues that this approach is both simplistic and harmful. Often the chain starts at the most convenient root cause, and usually only one root cause is identified. In addition, systemic factors that allowed the events to occur are often ignored. Leveson cites the Bhopal chemical accident as an example of how systemic factors create overwhelming events.

In December 1984, the release of methyl isocyanate (MIC) from the Union Carbide chemical plant in Bhopal, India, caused at least 10,000 deaths and more than 200,000 injuries, making it the worst industrial accident in history. The proximate cause was assigning a relatively new worker to wash out some pipes and filters. The worker neglected to insert a special disk to prevent water from leaking into the MIC tank. Water flooded in and caused an explosion. A relief valve opened, and 40 tons of MIC was released into the air above a populated area.

What really caused the accident? Was it an inadvertent error by an inexperienced maintenance worker, or was it part of a sabotage plot, as Union Carbide later claimed? Leveson believes that neither explanation holds. The pipe-washing operation should have been supervised by the second-shift supervisor, but that position had been eliminated in a cost-cutting effort. Inserting the disk was not the job of the worker washing the pipes, because he was a low-level employee and should not have been entrusted with this important task. Further, contaminating MIC with

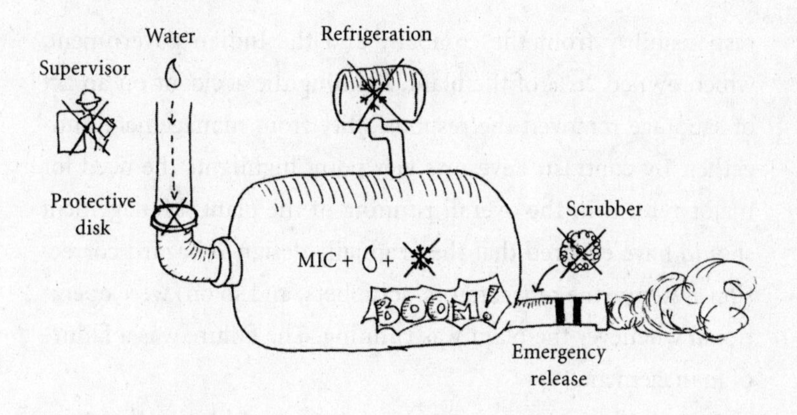

At the time of the accident in the chemical factory at Bhopal, most of the designed safety mechanisms (supervisor, refrigerator, and scrubber) were inoperative.

water was bad, but it should not have caused an explosion if the MIC were being kept refrigerated as called for by the design. The refrigerator, however, was not working. MIC itself should not have escaped up the flue, because a scrubber should have stopped it. But the scrubber was out of service.

In Leveson's analysis, the maintenance worker was only a minor and somewhat irrelevant player in the accident. Instead, safety degradation had occurred over time and without any particular single decision. A series of decisions had moved the plant slowly toward a situation in which any slight error would lead to a major accident. Although the Indian government blamed the accident on human error, Leveson would say the whole system was unsafe. Errors happen. The question is how to design and operate the system so that normal errors and failures do not lead to a catastrophic accident.

Blaming the accident on operator error removed most of the

responsibility from the company and the Indian government, which owned 26% of the plant. Blaming the accident on an act of sabotage removed the responsibility from management altogether. By contrast, Leveson's viewpoint highlights the need for major reform in the overall running of the plant. Management should have ensured that the originally designed hazard correction systems (the refrigerators, scrubbers, and so on) were operational whenever the plant was running. The failure was a failure of management.

System theory treats safety as a control problem rather than a failure problem. You control safety by enforcing constraints on the behavior of the system and its components. Control, in turn, depends on the notion of feedback. Think about how you deal with an unfamiliar hotel shower. You rotate the temperature dial to about the right level and then feel the water, adjust the control, feel the water some more, and so on. Even if the dial is very different from any you have ever seen, you are able to set the shower to a comfortable temperature in a short time. The single most important feature of feedback-based control is that the sensor (your hand) must be accurate while the control need only be *monotonic*—turning the dial clockwise always has one outcome: making the water cooler.

Effective control requires an accurate view of the existing state of the process, as well as the ability to exercise control. The Bhopal plant had inadequate control over the pipe-washing operation—the missing supervisor should have checked for safety. The plant could not exercise control over the released MIC, because of the inactive scrubbers. Meanwhile, the nearby population was never alerted to the possibility of a problem or even given instructions about elementary countermeasures that could be taken to prevent injury (a wet towel over the face). These are only a few of the problems. To truly understand why this accident occurred, we need to look beyond the plant and include the management oversight by Union Carbide and the Indian government. This broader perspective can help prevent similar future accidents.

In her studies of accidents, Leveson has noticed common features. First, there is usually *operational degradation* away from safety. Nothing bad happens for so long that safety personnel become complacent, management cuts budgets, and maintenance becomes shabby. Bhopal presents many examples, but there is nothing special about India. In southern California, for example, heavy rains often cause flooding. Because rain is infrequent in that part of the country, towns neglect to clear the debris from storm drains, making flooding much more likely when it does rain.

Second, accidents often happen because of *asynchronous improvement.* Suppose someone upgrades software (an optimization for that person) and then sends you a document using a new version of a word processor that is incompatible with your version. You are not able to read it, and communication is no longer possible. Sometimes a local optimization leads to a senseless loss of life and property. For example, in 1994 a pair of US fighter planes shot down a US helicopter. The fighters had acquired an updated communication system, but the helicopter still had the old system. The helicopter pilots never heard the warning calls from the fighters.

Third, accidents can happen because of *uncoordinated and dysfunctional interactions* among system components. Separately, each component operates correctly, but together their actions lead to trouble. For example, a few years ago two planes collided in midair over southern Germany. The Swiss air traffic controller gave the correct commands to the pilots to avoid each other. At the same time, an onboard collision avoidance device also gave commands to the pilots. Everything would have been fine if the two pilots had each followed the same set of commands. However, one pilot followed the command given by the ground air traffic controller, and the other followed those of the onboard automated system.

These three problems—operational degradation, asynchronous evolution, and uncoordinated and dysfunctional interaction—are clearly interrelated. Most of these problems happen after a process has already been established. Seeking greater efficiency, individual actors reduce budgets (Bhopal) or optimize their own operations (fighter planes). The result is weakened control and accidents. It's turning the shower dial without putting your hand out to test the water temperature.

Leveson's philosophy of safety-driven design follows from these observations. She argues for designing safety into the system from the beginning, using basic system-engineering principles and the hazard analysis technique that she has developed. In her latest book, Leveson offers a simple example of a train's automated door system. The first step in the design process is to identify the system hazards. Some of the hazards include the following: (1) A person is hit by closing doors. (2) Someone falls from a moving train. (3) Passengers and staff are unable to escape from a harmful condition in the train compartment, such as a fire. These conditions are translated into constraints for the design of the automated doors and their control system: (1) An obstructed door must reopen to permit the removal of an

Rod Brooks in front of Marvin Minsky and a beta version of Brooks.
CATHY LAZERE

A younger Stoica in Moldovia.
ADRIAN STOICA

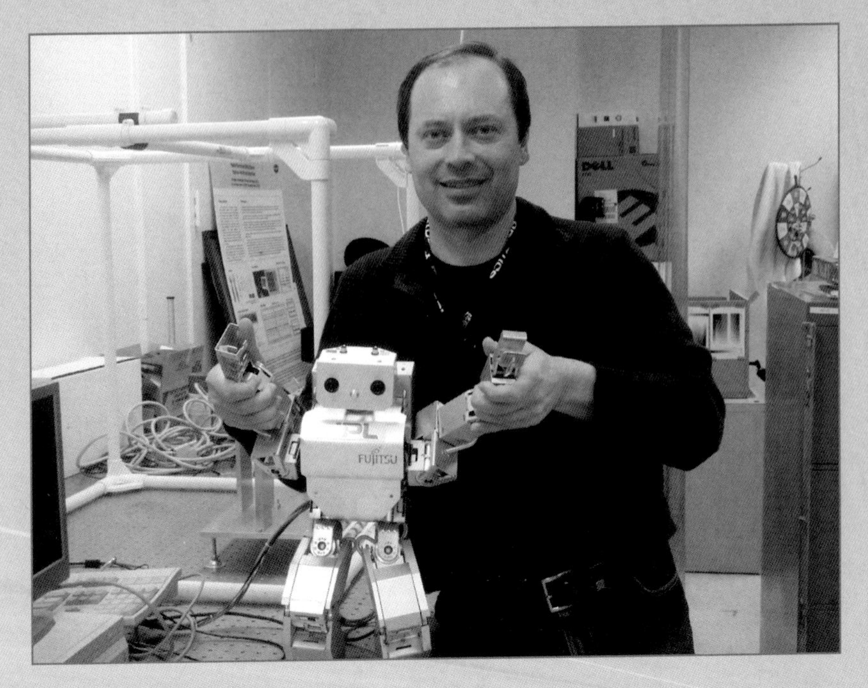

Adrian Stoica with a special friend. ADRIAN STOICA

A smiling Adrian Stoica with an evolvable circuit that hasn't learned how to smile—yet.

Lou Qualls in the nucleus of a carbon atom at Oak Ridge. CATHY LAZERE

The surfer dude and the tennis pro, Loveless and Bharambe, with a different kind of scoreboard. CATHY LAZERE

Ned Seeman on the edge of his desk at NYU. NED SEEMAN

Paul Rothemund in front of the DNA pool at Caltech. CATHY LAZERE

Steve Skiena and Steffen Mueller, one of his collaborators from the Department of Molecular Genetics and Microbiology, proudly display phages of a synthesized polio virus. Note: The scientists donned white coats but did not offer the photographer a Hazmat suit. CATHY LAZERE

Skiena takes his place among a pantheon of Indian gods.
CATHY LAZERE

Gerry Sussman, sporting a Nerd Pride pocket protector. CATHY LAZERE

Even the tables at Harvard are smart. Radhika Nagpal sits at one of them. ELIZA GRINNELL, HARVARD SCHOOL OF ENGINEERING AND APPLIED SCIENCES

Monty Denneau proving that a peak in Acadia National Park is scalable. MONTY DENNEAU

David Shaw with a graphite calculator. DAVID SHAW

The motherboard of an analog computer. CATHY LAZERE

Jonathan Mills and Bryce Himebaugh in their lab at the University of Indiana, Bloomington. CATHY LAZERE

Scott Aaronson and his charges in "The Complexity Zoo." CATHY LAZERE

Scott Aaronson's favorite diploma: his "get out of high school free" card. CATHY LAZERE

obstruction and then automatically reclose. (2) A door should open only when the train is stopped. (3) Means must be provided to open doors for emergency evacuation anywhere when the train is stopped.

These constraints may lead to various design decisions. For example, the first constraint could suggest the use of a sensor in a door similar to the one in an elevator. The second constraint might entail an interlock between a motion sensor and the door. The third constraint might combine the interlock with a "door open" instruction when an alarm is sounded. In effect, hazards give rise to constraints, which then give rise to design guidelines. The idea is to build in safety right from the start.

Making safety issues understandable at every level of the design, from the most abstract to the most concrete, may discourage operational management from optimizing costs at the risk of reducing safety. "Feedback and control" is a far better approach than assuming an absence of failures. After all, as folklore tells us, shit happens.

Leveson has applied system safety theory to the US Missile Defense Agency, which has constructed a giant interconnected system of systems. Some of the missile defense systems date back to the 1960s, including the cold war era's North American Aerospace Defense Command (NORAD). Others are more modern, such as newer radar and platforms. Leveson was called in to assess not just the reliability of the individual parts, but the safety of the integrated whole. "The analysis found so many feasible scenarios for inadvertent launchings that the deployment and testing phase of the new US missile defense system was delayed for six months while they fixed them," says Leveson.

Leveson also applies her system approach to a variety of civilian problems, including the safety of pharmaceuticals. For example, drug safety depends ultimately on chemistry and human physiology, but finding problems depends on appropriate testing. Relevant factors include the demographics and general health of the test population, the intended target population, and quality controls. The quality of testing, in turn, depends on the funding and power of the regulators. Regulatory power is affected by drug company lobbyist contributions. Lobbyists have more money to spend when their companies make large profits. So, as the drug industry becomes more profitable, lobbyists spend more, lawmakers reduce regulatory oversight, testing becomes more lenient, and drugs are brought to market more quickly, thus increasing profits and decreasing regulation even more. At some point, a poorly tested drug causes significant harm, leading to an outcry that ultimately leads to overregulation. To Leveson, the problem is not the rogue drug, but the fact that the quality of testing and lobbyist contributions to Congress are in the same control loop. The quality of testing should depend only on what is best for public health, not on who has the most powerful lobby.

Positive and negative feedback control loops can be found throughout nature. They keep organisms healthy and ecosystems in balance. System safety tries to do the same for human-built systems. A design that considers safety from the start can incorporate feedback control loops to prevent hazards. System safety leads to natural computing.

Part II

HARNESSING LIFESTUFF

IF SOMEONE MENTIONS THE WORD *COMPUTER*, YOU MIGHT PICture your laptop, your phone, or perhaps the hundreds of embedded computers in your car. And if you were asked what makes them tick, you would likely imagine electricity as the energy source and transistors as the switching elements. Despite current practice, however, computing has no necessary connection to electronics. After all, humans do it.*

Perhaps we do it so well because the fundamental biological building blocks—DNA, viruses, cells—all do it too. What if we could control those building blocks and make them compute for us? Is that even possible?

Ned Seeman, a professor of chemistry at New York University,

* When the English mathematician Alan Turing, the founding theorist of modern digital computing, conceived of his "Turing Machine," he used the term *computer* to evoke an image familiar to his contemporaries: a person (usually a woman) whose job it was to compute mathematical functions.

invented the basics of *DNA nanotechnology* out of frustration. He had been trying to crystallize proteins—a difficult and often tedious task. One day a colleague asked him to analyze a DNA Holliday junction, a short-lived four-way intersection structure that arises naturally in DNA. After understanding the junction, Seeman realized he could create arbitrary shapes using DNA, as opposed to its predominant linear double-stranded shape. He has since constructed structures at the molecular level, including DNA bricks and 10 nanometer–sized robots.

Seeman's technique exploits the fact that single strands of DNA will bind to one another if they can line up the letters of the DNA alphabet—the A of one strand is near the T of the other, and the C of one is near the G of the other. The net effect is that he can mix a few vials of single strands together and create a structure by a powerful technique known as *self-assembly.* Imagine being able to construct a geometric figure of very precise shape without manual intervention. That's self-assembly.

Seeman's seminal discoveries have led to a flurry of work involving the dynamics of DNA. **Paul Rothemund**, a young researcher at Caltech, has shown how to fold the 7,000-base DNA of an entire virus into the shape of a smiley face, again by self-assembly. By demonstrating that even large shapes can be made to self-assemble, he has set the stage for the micromanufacture of highly precise electronic circuits. As economically important as that endeavor could be, Rothemund's most exciting project is the insertion of large segments of specially shaped DNA into cells, leading to an entirely new way to study and maybe even to modify cell biology for medical therapies.

While Rothemund uses a virus as feeding stock for DNA self-assembly, **Steve Skiena** and his biochemical colleagues at SUNY

Stony Brook have modified a real poliovirus with the hope of creating vaccines. By injecting a bit of weakened virus into a subject, vaccines kick-start an immune response without killing the infected person. The body "remembers" the response, so if a stronger version of the same virus appears later, the immune system can recognize and destroy it. Finding weak viruses is difficult when done by trial and error, as Jonas Salk discovered in the 1950s when he tested the poliovirus on monkeys. Skiena has *designed* viruses out of the DNA code itself. The viruses multiply slowly enough to be weak, but they still trigger the appropriate immune response. Skiena's technique exploits asymmetries in the genetic code that can slow down the construction of viral proteins.

DNA can self-assemble, and viruses can hook into cells. But bacteria and multi-cellular organisms are even more talented—they can reproduce and self-organize. These capabilities offer the possibility of harnessing millions of cells at the bacterial level to perform a massive computation. Using many processors together is known as *parallel processing*, but this is parallelism beyond anything previously constructed in silicon, with millions or even billions of computing elements. **Gerald Sussman**, inspired by the ability of cells collectively to form an embryo, thinks that cells could be corralled to improve on nature. He imagines scenarios in which bacteria construct wood without flaws, adjust the viscosity of lubricating oils, or repair failures in our bodies.

One of Sussman's former students, **Radhika Nagpal**, has worked out a way to program cell-like objects. The challenge is to make fragile, asynchronous cells that talk only to their immediate neighbors to achieve global goals. Nagpal's method employs the idea of chemical diffusion. If a chemical signal starts at a particular location in a group of tightly packed cells of the same type, then the signal moves outward in all directions at nearly the same speed. Chemical diffusion suggests a way to coordinate millions of fragile biological cells, micromechanical devices, or tiny robots. Instructions would diffuse through a cellular medium at roughly uniform speed in all directions, even though individual cells themselves vary in their speed and some of them may die. The global instructions can cause millions of cells to coordinate and fold into precise shapes, maintain balance, or even build something new.

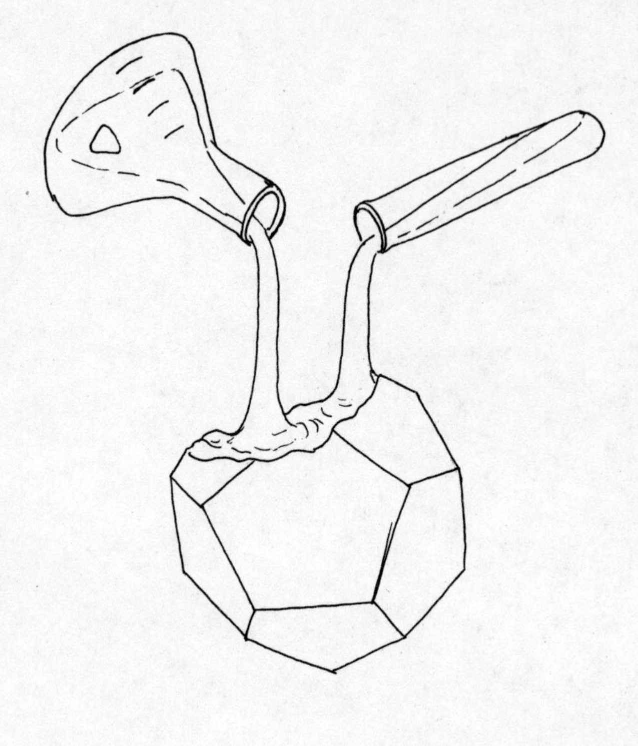

■ ■ ■ ■

I've always regarded the DNA sequence as a very long four-letter word. It's a one-dimensional object and I'm not a one-dimensional guy.

— *Ned Seeman*

Chapter 6

NED SEEMAN

. . . .

At the Edge of Life

NED SEEMAN BEGAN HIS RESEARCH CAREER AS A *CRYSTALLOGRAPHER*—someone who tries to understand the underlying, repeating patterns of molecular structures through X-ray diffraction. Over the years, crystallographers have made central contributions to our understanding of materials science and biology. In the 1880s, Pierre Curie and his older brother, Jacques, showed that deforming crystals could give rise to an electrical potential. The size of the effect depended on the type and quality of the crystal. This finding formed the basis of modern semiconductors, which now require crystals of excruciating perfection. In biology, crystallographers and their data led to an understanding of the fundamental components of life. In the 1950s, Rosalind Franklin's X-ray diffraction data for DNA helped guide Watson and Crick to the double helix.

Seeman—no longer a pure crystallographer—and his students at New York University spend most of their time devising chemical protocols to induce DNA to form molecular sculptures and robots. In the future, these sculptures could form a scaffold for the construction of nanoelectronic devices. Or they could morph into tiny robots creating nano-scale factories. Currently, about 35 labs around the world are working on DNA nanotechnology. Seeman started the whole thing in the 1980s in large part because of an extreme reaction to what he calls "the tedium and stupidity of the crystallization experiment."

Seeman was born in Chicago in 1945. His mother coined his first name, "Nadrian," after Seeman's paternal grandfather, Nathan. "He hated his name, so she made up a name I could hate," says Seeman. (Seeman uses "Nadrian" for his publications but prefers being called Ned.) In high school in Highland Park, a suburb of Chicago, Seeman was inspired by a biology teacher who explained the subject as a physical, chemical phenomenon. The section on genetics caught Seeman's attention. "From then on my interest was in that real boundary between living and non-living systems—the edge of life," he recalls. Seeman's father, a furrier, wanted his son to become a physician, so that's what Seeman planned when he went off to the University of Chicago. But after witnessing the long suffering of his dying grandmother and the medical world's inability to save her, the young Seeman decided that medicine wasn't for him.

Instead, Seeman studied biochemistry. At the University of Chicago, in the mid-1960s, this meant measuring metabolic reactions of small molecules. "*Unbelievably* boring!!" recalls Seeman. He decided he preferred disciplines with more math and entered a crystallography program at the University of Pittsburgh in 1966. "With crystallography, everything clicked," Seeman says. "There was computing and three dimensions in crystallography and a certain amount of math and physics that was easy for me to understand. And it turned out I was pretty good at it."

In 1970, Seeman finished his doctorate in crystallography/biochemistry. He solved his first nucleic acid structure (for an RNA molecule) in a postdoc position at Columbia (see the box "How to Solve a Crystal Structure"). Seeman continued working on the crystallography of nucleic acid components at Alex Rich's lab at MIT. "It was a hard crystal structure, the ideal project for me," Seeman remembers.

When Seeman applied to be an assistant professor of crystallography, the job market was poor. Crystallography was expensive; no school wanted more than one crystallographer. In the year of his application, a hundred crystallographers were looking for work and only six got jobs. Seeman was hired for what was arguably one of the better jobs. "But it was a scientific death sentence to be at SUNY Albany, where I spent 3,983 days, which I usually round off to 4,000," he says.

In those days, every crystallographer would choose a protein and work on it for years and years; that was the kind of commitment required given the state of the technology. Seeman didn't like proteins as much as DNA and RNA—nucleic acids. But even though nucleic acids had some logic to them, Seeman couldn't do the "voodoo" part of crystallography: grow a good-sized crystal. "It looked like I was going to wind up down in New York City as a taxi driver," Seeman recalls.

Then one day his luck changed. Bruce Robinson, a postdoc down the hall from Seeman at SUNY Albany, came to Seeman with a request from Robinson's boss, Leonard Lerman. Lerman wanted Seeman to look at a DNA structure called a *Holliday junction.* (Lerman was famous for his discovery of *DNA intercalation*, a process by which non-DNA molecules can wrap themselves into DNA.) A Holliday junction is a four-stranded

How to Solve a Crystal Structure

To grasp what Seeman accomplished in solving the structure of an RNA molecule you have to understand the basics of solving crystal structures. X-ray crystallography makes use of the fact that X-rays will be diffracted (reflected in different directions) by the electron clouds in crystals. The pattern of diffraction gives information about the density of the electrons. X-ray data also yields the amplitudes of the complex Fourier components, meaning that the data mathematically describes the repeating patterns of the crystal structure. But the data doesn't include the phases (shifts in the repeating patterns). "Solving a crystal structure" means working out the phases. With both the amplitudes and the phases, you can reconstruct the crystal structure. That is, you can figure out the positions of all the atoms.

When Seeman began his work, a structure was considered hard if it had a lot of roughly equal atoms. The sequence he crystallized (ApU) had 95 (including solvent) non-hydrogen atoms. (Hydrogen is virtually invisible to the X-rays.) Two of the non-hydrogen atoms were phosphorus (15 electrons), and the rest were first-row atoms with 10 or fewer electrons. So it was a tough structure. The term ***first-row*** comes from their position in the periodic table. If there are one or two very heavy atoms, such as iodine (53 electrons), it is pretty easy to figure out where they are, because an atomic signal is proportional to the square of the number of electrons. From there it is easy to work out the rest of the structure.

The sequence that Seeman crystallized was ApU. There

were two molecules in the asymmetric unit, and the coordinates of every atom in each molecule had to be determined. The sequence consisted of one phosphate backbone between two sugar-base units. There were thus two A-U base pairs in the system. This was the largest unit that had been studied crystallographically at the time. About 20 previous attempts on similar molecules had all failed.

structure that resembles a four-way road intersection. In nature, stable DNA has a linear topology. In effect, Lerman and Robinson were asking Seeman for help in building a model to explain how the nonlinear Holliday junction appears and disappears. The collaboration went well, but Seeman was interested in more than just observing that the Holliday junction changes from one configuration to the other because of the symmetry in the structure (see figure).

In a conversation with Kathy McDonough, one of his undergraduate students, Seeman realized that his team might be able to test some of the hypotheses they had generated by the modeling. He reasoned that if they made synthetic DNA, they could get rid of the symmetry of a Holliday junction and thereby create a stable junction. A stable, nonlinear DNA structure had never been seen before. That was the start of Seeman's work in DNA nanotechnology.

Seeman realized that four-armed branch structures were just the beginning. If they could make a stable branch structure, they

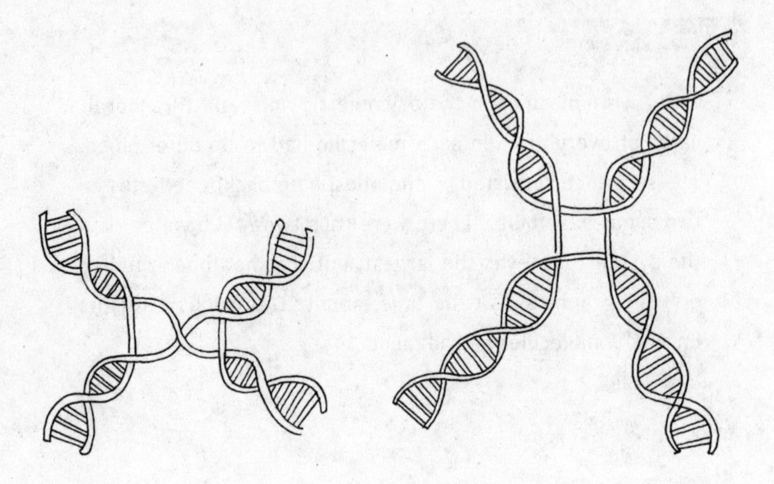

Two possible configurations of a four-way DNA intersection. In the first case, the top right and the bottom left helices are attached. In the second case, the top right and the top left are linked. The Watson-Crick attractions are similar in the two cases, so either state can occur.

could increase the number of arms. In September of 1980, Seeman went to the campus pub to think about six-arm junctions. The Dutch artist M. C. Escher's *Depth* came to mind. Escher's print shows a multi-dimensional school of fish, each with six propeller-like fins. "I realized that the flying fish were just like six-arm junctions," Seeman recalls. "The important thing was not just that the fish in the Escher picture are topologically six-arm junctions, but that the fish are arranged like the molecules in a molecular crystal. There is periodicity front to back, top to bottom, left to right." He thought he could use this idea to organize DNA into crystals.

In 1980, synthesizing DNA was laborious. "You have a molecule, and you don't know which interactions will hold the crystal together. Even once it's done, you don't even know why the thing has crystallized the way it has and not some other way," explains Seeman. But he decided that the goal—*rational crystallization*—was worth it. Rationally designed DNA crystallization depends on "Watson-Crick pairing": nucleotide A (adenine) wants to pair with T (thymine), and G (guanine) wants to pair with C (cytosine). As a result, it is possible to design a molecule and know what it is likely to look like when you X-ray it. That's the theory anyway. Then there's actual practice.

First Seeman had to learn to synthesize DNA in sufficient quantity. This task took him three years in the early 1980s. Next

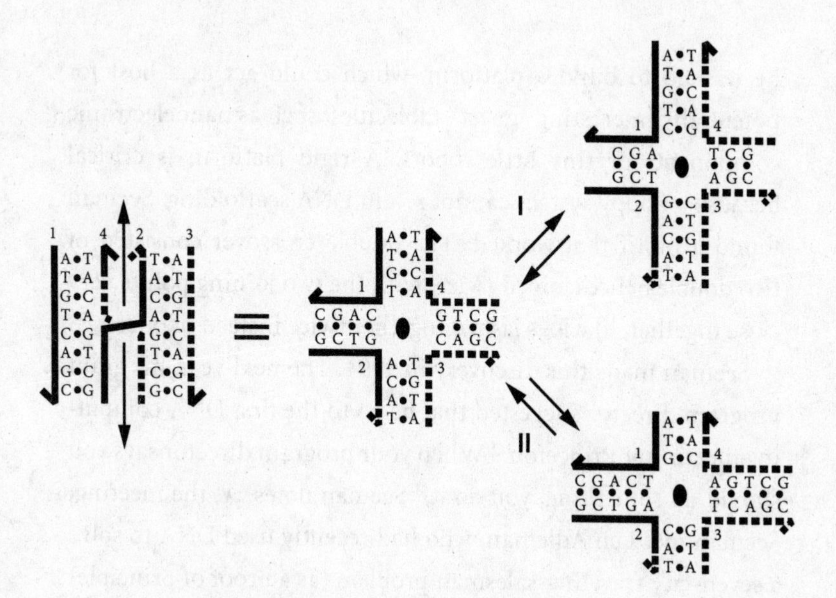

Some of the ways that strands of DNA may escape linearity. Watson-Crick pairing (A binds with T, and C binds with G) can create lots of different shapes. Courtesy of Ned Seeman.

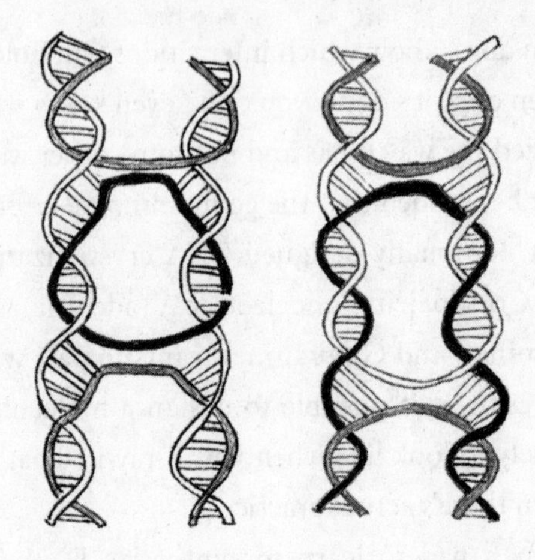

Two forms of the double-crossover motifs that Seeman developed as stable platforms for DNA constructions.

he wanted to build a platform, which could act as a host for potentially interesting "guest" molecules, such as nanoelectronic components or tiny little robots. A rigid platform is critical because a floppy system cannot retain DNA scaffolding. Seeman found a motif that worked. The double-crossover consisted of two double helices joined twice, with the two joining points very close together, like logs lashed together twice instead of once.

Seeman made this discovery in 1994. The next year, his grant program director suggested that he go to the first DNA computing meeting at Princeton. "When your program director says you should do something, you do it," Seeman notes. At the meeting, Seeman met Len Adleman, who had recently used DNA to solve a seven-city traveling-salesman problem (as a proof of principle) (see the box "Adleman's DNA and the Hamiltonian Path," on page 109). He also came across two young Caltech students—

Erik Winfree and Paul Rothemund (see the next chapter, on Rothemund).

Seeman and Winfree worked on extending the double crossover to a two-dimensional array that contained programmably spaced stripes. Next, Seeman's lab built a two-state nanomechanical device. More recently, Seeman made a little robot arm that flips back and forth in the array. Putting two of these together yields a double-armed robot that might pick up other pieces of DNA as part of an assembly process.

Once the fundamental components work, all kinds of possibilities open up. One possibility is to produce designer DNA-based drugs, discovering how well different drugs interact with potential drug receptors by testing their effectiveness and side effects. That process requires high-resolution crystals, and the laboratory issues remain challenging. Seeman is also interested in getting DNA to create a weaving pattern that could result in a kind of lightweight chain mail bulletproof vest, among other applications.

Seeman's work takes patience and a long-term vision. He has had students nearly crack from the repetitiveness of the lab work. Others have thrived. In the meantime, publicity has elevated nanotechnologists to celebrities in what Seeman describes as a "nanocult." At his first Nanomeeting in Seattle, Seeman and his colleague Bruce Robinson proposed a DNA-scaffolded nanoelectronic system. Many of the attendees were social scientists, science fiction writers, and software developers. "They were all culty people," says Seeman. At a party one night a young woman asked him, "Isn't it a pity that we're the last generation to die?" He replied, "Huh?"

Seeman sees a gap between people who conceptualize the implications of the science and those who live in the laboratory and endure the daily routine of coming in and failing. He sees part of his job as training students to deal with failure psychologically: "You walk into the lab and the odds are everything screws up." He also draws a distinction between what chemists and computer scientists experience. If a programmer knows the problem, eventually he will figure it out. In a chemistry lab, by contrast, what you want to do might just not be possible, given your current approach. Many things can go wrong, but overcoming the barriers can lead to enormous scientific progress, as in Seeman's case.

Seeman even proposes a rather morbid thought experiment to determine the value of a scientist's long-term accomplishment. He calls it the "abortion test." "Imagine that you were aborted," he says. "What don't we know, and do we care? Ultimately, this is about what is going to be written on your figurative tombstone. What did you do with your life, and what is the point of it? And did you do anything that ultimately has any impact?"

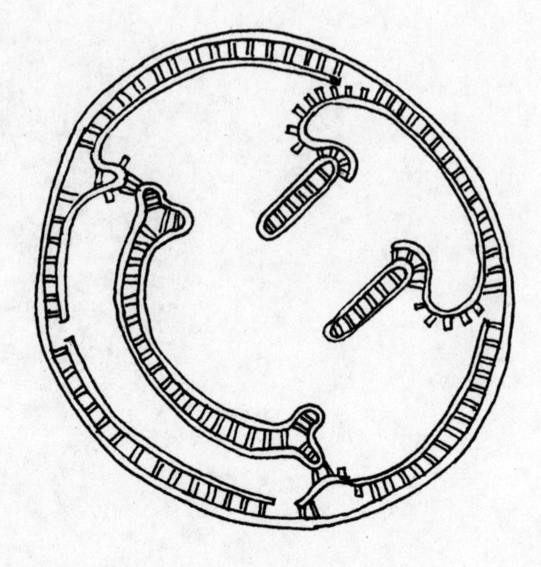

▪ ▪ ▪ ▪

DNA origami was a leap of faith. I had five other projects going at the same time. I started out small—I didn't want to waste too much money. But it was something that I thought was such a high-value target, that if it worked out it would be great.

— *Paul Rothemund*

Chapter 7

PAUL ROTHEMUND

▪ ▪ ▪ ▪

Lifestuff Imitates Art

ONCE NED SEEMAN HAD SHOWN HOW TO MANIPULATE DNA USING Watson-Crick complementarity, it was only a matter of time before the idea caught on and designers built structures upon the microscopic scaffolding. New York's Museum of Modern Art has already exhibited the DNA origami of Paul Rothemund, who is the recipient of a Feynman Prize in Nanotechnology and a MacArthur Fellowship. And he's just getting started.

Born in 1972, Paul (Wilhelm Karl) Rothemund was named after his chemist grandfather. His father, Max, is an ophthalmologist and his mother, Judith, a photographer. Rothemund grew up in Laconia, New Hampshire, a small town of 16,000 people about ninety miles north of Boston. He enjoyed tramping around the natural beauty of the area's many lakes and mountains. He loved to catch snakes and take them to school in his backpack, which got him in trouble. Nevertheless, Rothemund's elementary school principal, Irene Wright, encouraged his interest in science and technology. In the early 1980s the school had purchased

TRS-80 computers, but the teachers didn't know what to do with them. "They pulled me out of class in the second grade and put me in a little room with computers and the manuals and said, 'Figure out what to do with these,'" Rothemund remembers.

Rothemund thought about becoming a field biologist. But he says he realized he was about a hundred years too late to make a contribution in that area. Then he took up scuba diving and "wanted to be Jacques Cousteau." One day at age eighteen, while he was standing in the shower, his lung collapsed, ending any possible career as an underwater photographer. Maybe it was for the best. He had always felt he was much more of an engineering type, because he enjoys complete understanding. "It's heartbreaking to me when you have a complicated system and you think you understand something, and you find out later there is some hidden actor or component you didn't know about. By contrast, with computer and engineering systems," he notes, "you know all the pieces that you put in."

Rothemund devoured Richard Feynman's books and was a fan of the movie *Real Genius*, a 1985 satire set at a fictitious university called Pacific Tech, closely modeled after the California Institute of Technology (Caltech). The film includes some memorable lines, such as, "Would you be prepared if gravity reversed itself? The only thing I can't figure out is how to keep the change in my pockets. I've got it. Nudity." Primed by the movie and the well-known history of Richard Feynman's illustrious teaching at Caltech, Rothemund became an undergraduate there and studied biology. During the summers he did chemistry research, and he ended up receiving a combined bachelor's degree in computer science/engineering as well as biology.

One of Rothemund's computer science professors was Jan L. A.

van de Snepscheut, a student of the pioneering Dutch computer scientist Edsger Dijkstra.* Snepscheut instilled in his students an appreciation for the physics of computation. "The spirit was that computation is a universal phenomenon and can be performed by almost any substrate you can think of, from billiard balls to Tinkertoys to DNA molecules," Rothemund recalls.

In 1992, Rothemund got even more excited by that notion when he read a paper written ten years earlier by Charlie Bennett of IBM Research. Bennett talked about computing with DNA. Snepscheut encouraged Rothemund by suggesting that someone with a knowledge of biology, chemistry, and computers could figure out how to build a computer out of DNA. Later, for his term project in a class taught by Yaser Abu-Mostafa, Rothemund tried to meet this challenge. He proposed a way to build a computer that would cut and paste DNA strands and thereby simulate the action of that most classical model of computation, a universal Turing machine. After graduation, he presented the idea to some Caltech professors. "They uniformly thought I was crazy. They didn't know where to send me or what to suggest," remembers Rothemund.

Discouraged by the tepid reception, Rothemund says he "languished." He worked in geobiology for the better part of a year. Then, in the fall of 1994, Len Adleman published a seminal paper demonstrating DNA computation. While Rothemund had proposed a way to perform any possible computation, Adleman had solved only a small instance of the well-known Hamiltonian path problem, but with great elegance.

* A short biography of Edsger Dijkstra can be found in our earlier book, *Out of Their Minds: The Lives and Discoveries of 15 Great Computer Scientists.*

*We'll encode various symbols by sequences of DNA bases (A, C, T, G). For example, for city X the symbol representing "To X" (**toX**) will be ATT, and the symbol representing "From X" (**fromX**) will be GTCT. Then **toX'** will be the complement of **toX**—TAA; and **fromX'** will be CAGA. City X will be represented by **toX** and **fromX**; and an edge from city Y to city X will be represented by the "leg" strand **fromY'/toX'**.*

The Hamiltonian path problem is to find a route through a collection of cities (*N* cities) that starts in a home city, ends in a final city, and visits each city exactly once along the way. Suppose each leg in the route is a one-way flight from, say, city X to city Y. Adleman computed the solution to this problem by cleverly encoding all possible solutions using DNA, and then using common lab techniques to "fish out" a correct solution (for more details, see the box "Adleman's DNA and the Hamiltonian Path").

Adleman chose the Hamiltonian path problem because the only known guaranteed solution requires looking at all possible routes through all the cities. The natural parallelism of DNA computing offers a potentially inexpensive way to try all possible routes. Unfortunately, the DNA technique does not guarantee finding all possible routes for larger-scale Hamiltonian path problems. A big enough problem (for example, covering 1,000 cities) would mean that the number of routes exceeds the num-

Adleman's DNA and the Hamiltonian Path

To start the process, Adleman represented cities and legs between cities as single strands of DNA so that each "leg strand" had a left half that could bind the DNA for departure city X (via Watson-Crick base pairing, A binds to T and C binds to G) and a right half that could bind the DNA for an arrival city Y. When all city and leg strands were mixed together, the leg strands held the city strands together in chains of various lengths—for example, from city Y to city X to city Z, if such a route existed.

Adleman then increased the number of DNA molecules that began and ended in the home and final cities. He used a DNA reproduction technique called the ***polymerase chain reaction***, which amplifies a single chain thousands or even millions of times.

He located strands whose length indicated that they had gone through exactly ***N*** cities, by using another workhorse technique, called ***gel electrophoresis***, which uses an electric current to separate single- and double-stranded DNA by length. There was still the possibility that some of those strands included the same city twice while missing other cities. So Adleman passed his solution through a series of filters that used magnetic balls coated with DNA complementary to the DNA for each city. This procedure captured strands that had visited every different city at least once. Any molecules surviving this treatment were Hamiltonian paths.

ber of atoms in the known universe—you would need a really large pot of DNA for that experiment.

Adleman's paper generated enormous excitement and caught the attention of Erik Winfree, a graduate student who had also taken Abu-Mostafa's class. Having seen Rothemund's presentation, Winfree asked to meet Rothemund at the Red Door Café near Caltech to discuss DNA computing. At their second meeting, Winfree presented the idea of algorithmic self-assembly, simulating cellular automata by crystallizing little DNA bricks into aperiodic crystals. Rothemund wrote up his DNA Turing machine, and Winfree wrote up his DNA cellular automata. The two young scientists paid their own way to Princeton to present their papers at the first Conference on DNA Computing in 1995. There, Winfree and Rothemund went out to dinner with Len Adleman and Ned Seeman (see the preceding chapter, on Seeman).

Rothemund became a graduate student in Adleman's lab at the University of Southern California, working there for six years. Winfree collaborated with Seeman but stayed at Caltech and became a professor. The two friends maintained an ongoing collaboration, which continued as Rothemund returned to Caltech for a postdoc with Winfree. In Rothemund's doctoral-thesis dedication, he refers to Winfree as "an epic hero," while contrasting himself as "more of a forest wild man."

At Caltech, Rothemund is now a senior research associate with his own lab. Over the years, the fields of DNA computing and structural DNA nanotechnology have essentially merged. The DNA bricks required for algorithmic self-assembly were invented by Seeman, and it took a collaboration between Winfree, a computer scientist, and Seeman, a crystallographer, to create the first two-dimensional DNA crystals from these bricks. The art of sculpting such new geometric structures had always seemed like black magic to Rothemund; but by 2002, Rothemund had started to create long DNA nanotubes from the bricks and was taking a detour from DNA computation through DNA structure.

DNA origami came out of a conversation with then–UCLA graduate student Jonathan Kelley, who challenged Rothemund by asking if he could make large repeating patterns and algorithmic patterns—could he beat the world record for the largest non-repeating arbitrary structure? At the time, Seeman held the record—for a roughly 2,000-base stick-figure polyhedron known as a "truncated octahedron." The structure had 49 strands assembled over seven steps—it was a complicated feat, one that seemed hard to overcome using the same approach.

To make large, non-repeating structures requires lots of unique "chemical information"—lots of unique DNA sequences. To put them together is a lot like baking a cake: you have to make sure you get the ratios of all the ingredients right. Rothemund hit upon the idea of using a single long strand, so that success or failure would depend mostly on a single main ingredient. Rothemund knew that a variant of the virus M13 (M13mp18) used a circular single strand of DNA 7,249 bases long for its genome (recall that each base can be adenine, cytosine, thymine, or guanine). "The nice thing is that viruses really want to live. They have elaborate machinery for making sure that anytime an instance of the virus is copied, it's copied with very high fidelity. It's the most widely available, pure, long single strand of DNA," says Rothemund.

Imagine now that you are given a long strand of DNA and you want it to bend back and forth to fill out a given shape. You might lay it down in that shape, but DNA is in solution, so assume that your string is effectively weightless—it will writhe and twist all over. You might think of tying or stapling one section of the string to another section to ensure that they remain close to one another. To do that with the virus, Rothemund proposed the notion of DNA "staples."

Each staple is 20–40 nucleotides long, about 30 on average. For each long strand of DNA from the M13 virus, Rothemund mixed in about 200 short synthetic DNA staples. He added a little bit of salt water buffer, which the DNA needs to form a double helix.

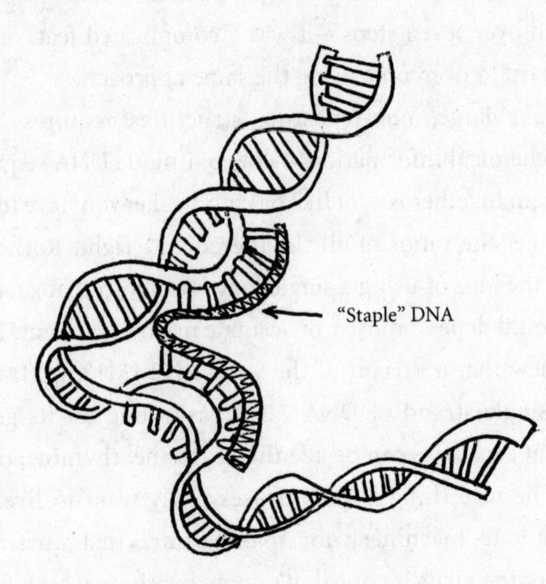

A small strand (the "staple") can hold parts of a larger strand (the virus in this case) together.

Then he heated it up to near boiling to unfold the virus and cooled it down to room temperature over the course of 2 hours to allow the staples to bring parts of the virus into the target fold. And he was done. Says Rothemund of his recipe, "It's very, very simple."

Conceptually, it is good to think of each staple as having two halves—a right half and a left half. The left half of the staple binds the virus strand in one place, and the right half binds the virus strand at a distant location. Staples bind to the correct pairs of spots on the virus thanks to strand displacement (see the box "DNA Mating Rituals"). The net effect is to bring the two distant locations together. Rothemund used about 200 different staples to create smiley faces at an angstrom scale. A meter is 10 billion angstroms, so we are talking about really small distances.

Rothemund has now made this process amazingly easy. A DNA artist draws the shape—a task that requires different strands of the virus (for example, X and Y) to lie close to one another. A computer program computes the Watson-Crick complement of X and Y and creates a staple sequence consisting of the two complements linked end to end. Altogether, a complex shape might need a few hundred staple strands. He e-mails the sequences of letters to a commercial supplier, who uses a DNA synthesizer to create the staples. They arrive in the mail. Each short strand and the virus itself comes in a separate tube. A robot mixes them together and adds a little bit of salt water, and the solution is then heated and slowly cooled. Each drop has 50 billion instances of the artist's vision (see figure). If the DNA synthesizer were on-site, the time from concept to molecules in the tube would be about 6 hours.

DNA Mating Rituals

A strand of DNA consists, abstractly, of a sequence of the letters A, C, T, or G. Two strands can join together if they have a stretch of "Watson-Crick complementary letters": A is complementary to T, and C is complementary to G. The complementary stretches from the different strands will stick together, but non-complementary stretches may flop around.

For example, consider a strand X = AACCCTTTT and another strand Y = AAATGGGATTTTTTTT. We can use boldface to highlight regions where each has a sequence of letters complementary to the other—for example, a pair of five-letter regions (A**ACCCT**TTT and AAA**TGGGA**TTTTTTTT) or a pair of three-letter regions (AACCCT**TTT** and **AAA**TGGGATTTTTTTT).

Because of the thermodynamics of the reaction, two strands will stick together for longer periods of time where the complementary matches are the longest. In a test tube, the strands would be bound by the five-letter regions much more often than by the three-letter regions. By carefully designing the DNA sequences, scientists can craft sequences that bind the way they desire almost all of the time.

Despite scientists' best designs, or in cases where it is more practical to use natural DNA sequences (the M13 virus for DNA origami, for instance), DNA strands sometimes bind incorrectly by long regions, and a simple two-state "bound or unbound" model for DNA predicts that they should be stuck in this "wrong" configuration for a long time, never reaching thermodynamic equilibrium. Fortunately, DNA has a way around this problem: ***strand displacement***. Suppose that a DNA designer wanted strand Y from our previous example to bind to strand Z = TTTACCCTAAAAAAAA, its perfect complement. Suppose further, however, that Y is already bound to X by a five-base region. Because Z binds to Y over a longer region, Z can sidle up to Y and displace X one base at a time until X falls off. That is, Z makes Y divorce X and run off with Z. Hollywood fans are familiar with this phenomenon.

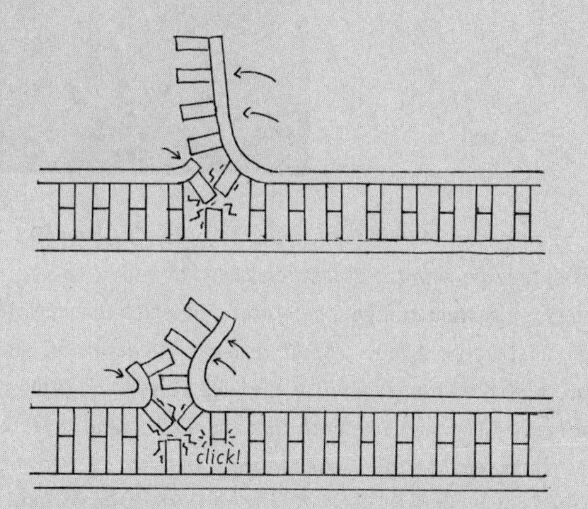

A longer match (coming from the right) will displace a shorter match in the vast majority of cases.

Bernie Yurke (of Lucent Bell Labs) first used strand displacement to demonstrate a ***DNA tweezer*** in which a series of strand displacement reactions caused the tweezers to open and close. Now, almost all DNA nanomachines and DNA computers are based on this principle.

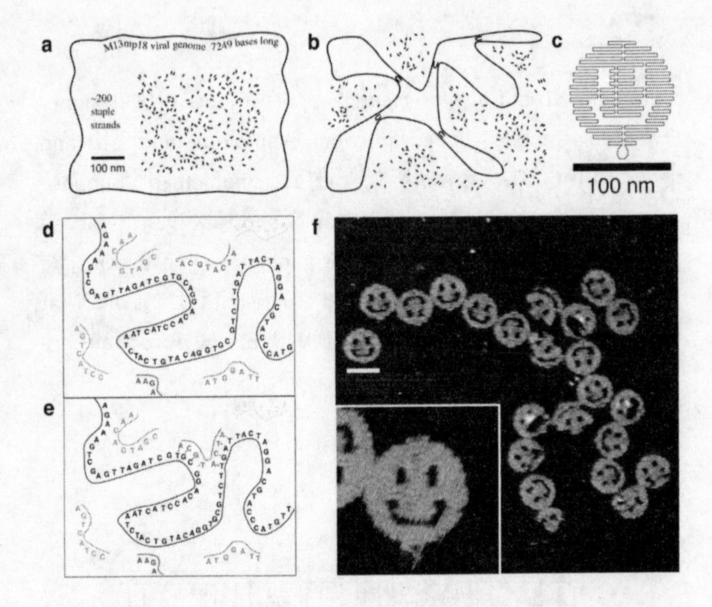

The folding of a DNA origami smiley face. (a) The long M13 viral strand, to scale with the 200 staples used to fold it. This cartoon depicts the system at high temperature, at which the long strand is an unstructured loop. (b) At a lower temperature, staples begin to bind distant regions of the long strand and bring them together. (c) The mazelike path that the long strand takes when the system reaches room temperature and folding is complete (staples not depicted). (d, e) A schematic close-up of how staples bind and fold the long strand; the ACGTACTA strand (upper middle) enforces a single constraint. Typically, staples are much longer than 8 bases (they average 30), and sometimes they bind the long strand in three places rather than just two as shown. (f) An atomic force microscope (AFM) image of about 20 DNA smiley faces stuck to a mica surface. The AFM takes an image of the origami by "feeling" the surface with a micromachined needle. Smiley faces appear to be delicate and sometimes break upon landing; about 70% of the smiley faces appear intact. About 1,000 smiley faces would fit across the width of a human hair.

If this sounds a lot more like programming than chemistry, Rothemund agrees. "Because computer scientists are good at this kind of thinking, I encourage them to join us as molecular programmers," urges Rothemund. While other researchers have made larger, as well as three-dimensional, origami structures using these principles, Rothemund wants to push the limits of size and complexity with new self-assembly principles. First, he wants to use the coarse geometric shape of the origami to create interlocking jigsaw puzzle pieces based on a peculiar property of DNA helices.

You can force the edge of a DNA origami to present rows of DNA helix ends, like a wall of Lincoln Logs that are cut off on their ends. If you make another DNA origami whose "Lincoln Log" edge has a shape that is complementary to the first one, the two origamis will stick together. That kind of jigsaw puzzle approach might permit the self-assembly of much larger structures. Rothemund eventually wants to use this method in combination with other techniques to make structures visible to the naked eye.

In other work, Rothemund and his collaborators are exploring how DNA origami can be used as a platform for the assembly of electronic components. Think of it: relying on the principle of pure self-assembly, mix a few tubes of DNA to create the origami, add some carbon nanotubes, burn away the DNA, and you've got a circuit. No etching, no clean rooms, no fancy fabrication. Of course, nanocircuits built in solution are of no use if they can't be transferred to a surface and integrated with conventional electronics. Along with IBM's Almaden Research Center, Rothemund is working on the problem of placing DNA origami at specific places on silicon surfaces.

Building breadboards for electronic circuits, while potentially very useful, neglects the ability of DNA origami to interact with other biomolecular entities. Rothemund is exploring *protein walkers*, proteins that make your muscles move by traversing tracks built by other proteins: "With DNA origami, we can make a DNA track, place binding sites for the protein walker wherever we want, and ask how the protein walker moves." Rothemund suggests that scientists will be able to marry DNA structures with protein motors, in order to achieve control over structure, computation, function, and movement. "When we can recapitulate and engineer it," Rothemund says, "we will have arrived at our goal: living nanotechnology."

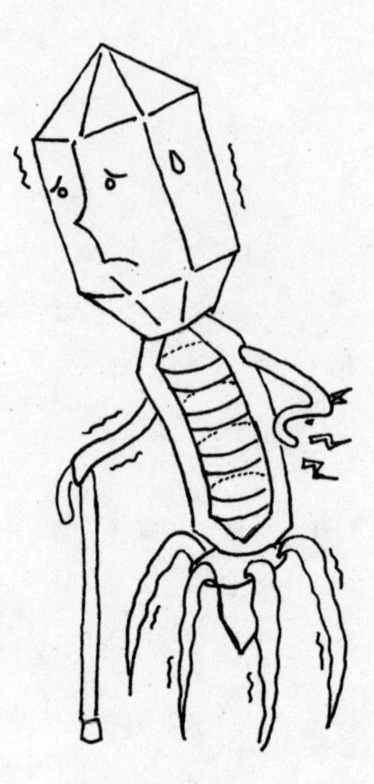

■ ■ ■ ■

To study a biological bug, we reprogram it to figure out how it works. Or we can optimize it in some way to make it more useful. This is very much a computer science kind of thinking.

— *Steve Skiena*

Chapter 8

STEVE SKIENA

▪ ▪ ▪ ▪

Programming Bugs

A VIRUS, AT ITS CORE, IS JUST A SEQUENCE OF DNA OR RNA. BY itself, a virus is not a problem, but the proteins that the virus makes can cause big problems by attaching themselves in a home invasion of healthy cells. Steve Skiena and his colleagues at the State University of New York (SUNY) at Stony Brook design synthetic viruses in the hope of making better vaccines to fight viruses. It's a little frightening, we admit, but Skiena assures us that the positive outcome of having another weapon to fight disease outweighs the unlikely negatives of an act of bioterrorism or an uncontrolled accident.

To understand the biocomputational design ideas that Skiena uses, you need to know a little about biology. Proteins are the general workhorse of biological beings and the support matrix of bones and skin. They are made by a two-step translation process in which (1) a string of DNA is copied ("transcribed") to RNA, and (2) the RNA is "translated" into a protein. The word *translation* is appropriate because a special little biological machine called a *ribosome* assembles a protein by reading three letters of RNA and fetching the corresponding amino acid. Then the ribosome moves on to read the next three letters of RNA and attaches the next amino acid onto the previous one. A protein is simply a string of amino acids.

The DNA/RNA string is written in a four-letter alphabet (A, C, T, G; though the T is replaced by U in RNA),* so there are 64 possible three-letter sequences. These sequences (called *codons*) are used to specify the 20 amino acids. This means that several different codons of the original DNA may translate to, or "code for," the same amino acid. For example, GCT, GCC, GCA, and GCG all code for the simplest amino acid, alanine. Therefore, the DNA sequence GCTGCCGCAGCG would eventually translate to a sequence of four alanines.

Why would nature allow multiple ways of coding for the same amino acid? It turns out that certain codons of a given amino acid are favored over others. The favoritism is species-specific. For example, in humans the GCC triplet is used 40% of the time for the amino acid alanine, making it by far the most popular choice. The more frequent triplets/codons translate to amino acids faster than the less frequent ones. So, making a gene with less frequent codons means that the protein as a whole will be manufactured more slowly than otherwise.

If a virus variant makes its proteins more slowly, that variant may be a good vaccine. Injecting a slow-growing virus gives the human body enough time to recognize the virus and to build

* As discussed in the previous chapter, on Paul Rothemund, each of these letters represents a particular nucleotide: adenine (A), cytosine (C), thymine (T), guanine (G), or uracil (U).

antibodies against it. If the virus were to multiply more rapidly, it would cause sickness. Now you know the basics of virology. We'll soon see how computing gets involved.

Steve Skiena was born in 1961 and grew up in suburban New Jersey. His father began his career as a radio repairman working for the legendary audio inventor Avery Fisher, at a time when Fisher had only three employees. "Showing the Skiena business sense, my father left Avery Fisher, thinking that the future was television repair," says Skiena. After working as a TV repairman, Skiena's father became an engineer at Bell Labs, even without the benefit of a college education. Skiena's mother stayed at home to raise the kids when they were young. Later she managed the investment of funds for the local government.

Every year, the family went on vacation to Florida, where they regularly attended jai alai matches. Jai alai is a high-speed ball game that originated in the Basque country in Spain and southeastern France. Wicker baskets called cestas are used to fling balls against a stone wall. The balls achieve speeds of almost two hundred miles per hour—sort of like a game of squash on steroids. To the young Skiena, the big attraction of jai alai was that his father gave the kids a small amount of money to make bets on the games. Skiena was fascinated by calculating the odds. After following the advice of a local tout one year, he netted enough money to take the whole family out to dinner. From his start as a juvenile gambler, Skiena built a successful betting system for jai alai using a computer simulation that takes advantage of the knowledge of team strengths and the order of play for the eight teams in the league. He describes the theory in his book *Calculated Bets: Computers, Gambling, and Mathematical Modeling to Win.*

Though wagering on jai alai requires lots of data and ranking the probabilities of eight teams, other sports offered simpler betting possibilities. In 1977, as a 16-year-old, Skiena wrote a computer program called Clyde to predict football games. He even published a column with his picks for the local newspaper, the *New Brunswick Home News.* Clyde would try to pick the winner of each football game. "That I don't do this for a living tells you something about my track record," admits Skiena.

At the University of Virginia, Skiena studied engineering. When he graduated in 1983, he didn't see any jobs that appealed to him, so he opted for graduate school in computer science. He had to choose between Cornell and the University of Illinois at Urbana-Champaign. He decided Cornell was too theoretical and opted for Illinois. Ironically, he wrote his thesis on theoretical computational geometry. "It didn't involve a computer in any way except to type it up," he notes.

Skiena joined the computer science faculty at SUNY Stony Brook in 1988, and he has been there ever since. In 1991, he noticed a close relationship between a computational geometry problem he was studying and the process by which enzymes cut DNA. Skiena asked some colleagues in the biology department if there were any problems that might lend themselves to a collaboration. "The biologists told me I could help by installing the latest version of Windows on their machines. I ran out of the building and didn't come back for many years," Skiena says.

Until the mid twentieth century, biology relied mostly on observation and the accumulation of facts. According to a joke told by Russian mathematicians, science is either physics or bean counting. Russian mathematicians thought chemistry was similar to physics, but biology lay firmly in bean-counting territory. Each bean was an observation about a creature in a habitat. To a mathematician, accumulating facts holds no appeal.

In the early twentieth century, that "data curator" quality of biology started to change. The genetic studies of the nineteenth-century Austrian monk Gregor Mendel inspired some pioneering work on the linkage of traits. Most notable was the work of the American scientist Thomas Hunt Morgan. He won a Nobel Prize in 1933 for his study of inheritance in the fruit fly *Drosophila*. Morgan established that genes were carried by chromosomes and caused mutations, and that some genes were more closely linked together than others, depending on their position along the chromosome.

Even without understanding the actual structure of genetic material, biologists could extrapolate that the genes coding for trait X were likely to be physically closer to the genes coding for trait Y than to those for trait Z. The reasoning was simple: organisms that had X were overwhelmingly likely to have Y, whereas they might or might not have Z. Trait correlation permitted a primitive kind of mapping from traits to locations on a still-unknown genetic material.

In the mid-1950s, James Watson and Francis Crick changed everything. Building on the crystallography of Rosalind Franklin, they explained the structure of DNA. Using DNA analysis, scientists later validated most of the Morgan-style mapping. What followed were forty years of exploration into the effect of individual genes on organisms.

This productive work resulted in an extreme form of reductionism. A researcher would focus on a single gene, manipulate its behavior, and then observe its effects. A typical scientific presentation would feature the researcher's gene at the center of a chart with arrows showing its effects radiating out in all directions.

In the late 1980s, computers and instrumentation reached a level at which researchers could begin to look at entire genomes rather than individual genes. Computers and computer science soon became an essential tool for gene comparison and gene assembly. These developments led to the sequencing of the human genome; to large databases of the DNA sequences of different species; and, finally, to the design of living organisms, the subject of Skiena's work. His computer science background proved to be essential.

Bioinformatics, applying information technology to molecular dynamics, is largely about the analysis of DNA sequences. Skiena knew he could read sequences, but he wondered if he could create them. And the natural question (to a computer scientist) then became, What is the best sequence and how do you design it?

Most universities pay at least lip service to the idea that scientists should work with researchers from other disciplines, because the knowledge of the different disciplines can lead synergistically to new advances. Knowledge is only one of the benefits. Another benefit is perspective. Computer scientists care about methods—finding a better way to perform a particular function—storing and retrieving information efficiently, solving a differential equation, or providing a physically realistic computer graphic model. But unlike researchers in many other disciplines, computer scientists don't care about the data itself. This data indifference has a major selling point: the methods invented by computer scientists can be applied to a wide range of fields. Database systems, for example, can be used for commerce, art, and space exploration without significant change to the underlying methods. The danger is that a computer scientist may make an assumption that simply doesn't hold for the application at hand.

Aware of this problem, Skiena looked for a biological collaborator. His dream was to synthesize sequences of DNA having several thousand letters. At that time, constructing sequences of fewer than a hundred base pairs was the state of the art. Skiena looked far and wide but couldn't find a biologist who wanted to join him in this effort. One day in July of 2002, when he was on sabbatical in Padua, Italy, Skiena picked up the *International Herald Tribune*. On the front page was a picture of a scientist at Stony Brook whose office was only two hundred yards from his own. The scientist looked familiar. Skiena had met him a couple of times but didn't know what he was researching. It turned out it was Eckard Wimmer. Wimmer made the news because he had synthesized the poliovirus from scratch by taking its library sequence and designing short pieces of DNA that would self-assemble into the target. Wimmer had inserted a small number of changes in the code to prove that this was his poliovirus and not somebody else's. "In principle, there was no reason that you couldn't make a virus from scratch. The idea that someone was conjuring up a virus in a test tube was exciting," remembers Skiena. "Some people thought it sounded dangerous, but full genome synthesis opens the door to many exciting applications."

Getting scooped is part of the life of any scientist. The good ones recover quickly. Thinking about a problem in depth makes you become adept at solving a whole collection of related problems. Skiena thought his computer science perspective might be useful to Wimmer to help him optimize the virus in some way. Skiena now had an ideal collaborator. Wimmer could synthesize and experiment with viruses that were designed using Skiena's algorithms. The National Science Foundation and National Institutes of Health gave them grants to design weaker versions of the virus that would produce the same immune response as the real virus.

Immune response is governed largely by how your immune system recognizes viral proteins. If scientists encoded the virus so that the proteins were exactly the same as the original, then the attenuated (weakened) strain would provoke the same immune response as the original strain. The question became whether they could design a sequence that would code for the same proteins but not be as good for the virus; it would replicate more slowly so that the body would have a better chance to fight it off.

In other words, they wanted to create a vaccine. Of course, polio vaccines already existed. In one of the great successes of pre-genomic science, in the mid twentieth century Jonas Salk and then Albert Sabin had created polio vaccines. They both had to engage in a lot of monkey business (rhesus) before injecting people with an attenuated virus. "The Salks and Sabins of the world went through an enormous number of monkeys. They evolved strains that liked living in monkeys much better than they did living in people," notes Skiena. There was a catch, however. The sequences that evolved in monkeys were similar enough to those of human virulent strains that they could occasionally mutate back to virulence if they were injected into people. For that reason, the Salk approach was to grow the virus in the laboratory, allow it to create its proteins, kill the virus with formaldehyde, and then inject the proteins into people. The body would then create antibodies to the proteins.

In contrast to Salk's killed-virus approach, the Sabin vaccine was a live virus that was weak. Its advantage was that it provided some herd immunity. If I hadn't been vaccinated but you were, I might catch a case of your vaccine virus. And in the process I, too, would become vaccinated. But there was a catch here too. Because the disease-causing virus is only a slight variant of the vaccine strain, the vaccine itself can (very, very rarely) revert to a dangerous form of the virus. Health officials knew they were taking a risk with the live vaccine.

The Sabin vaccine finally eradicated polio from the Western world, but the virus still rages in parts of Asia and Africa. These days, the question is how to make poliovirus extinct. What holds for polio also holds for smallpox and many other dreaded diseases. To prepare defenses against new viral diseases (such as SARS) or accidental or deliberate outbreaks, it might become necessary to create a new vaccine in a hurry. Skiena had some ideas. "Passaging viruses through other animals is a slow and painful process. Could we instead design a weakened version, synthesize that, and thus avoid relying on random mutations?" he wondered.

Skiena thought of the genome as a program and reasoned that it shouldn't be too hard to make changes that would make a program run worse. Skiena and his team of virologists began looking at ways to design viruses that would be less efficient at reproducing. In their first paper together, Skiena's team members observed that if they replaced popular (more frequent) codons with unpopular ones, they would get viruses that would reproduce very slowly. Imagine a reverse high school contest with prizes for the slowest and the least likely to succeed and you will get the idea.

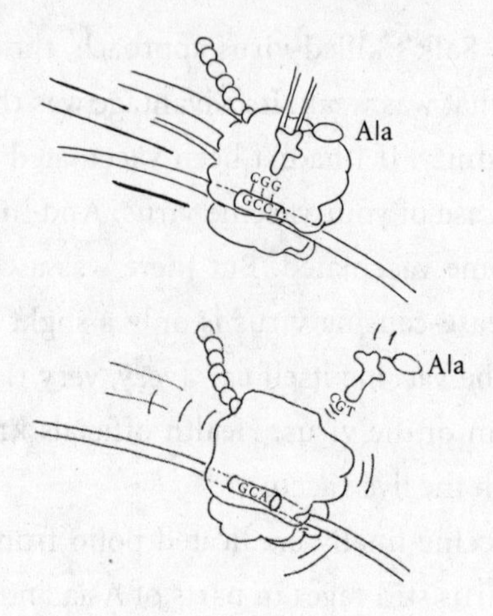

As we learned earlier in this chapter, in some organisms, such as humans, alanine (Ala) will be formed more quickly from GCC than from GCA, increasing the speed of production of the protein. So even though different nucleotide triplets may encode for the same amino acid, the efficiency of the encoding can vary.

Skiena's job was to design the sequence to use inefficient codons. His colleagues took it from there. His virology collaborators had been working with poliovirus for 30 years. They had plates of cells to measure the growth curves of the virus—how fast it kills cells. "By putting in unpopular codons, we designed a slow-growing virus," he explains.

Life science often proceeds from observation to application without completely passing through understanding. Aspirin, penicillin, and quinine water come immediately to mind as remedies that were used commonly before they were understood. So it is with the rare-codon strategy for designing vaccines. Even without understanding the "why" of organism-specific codon preferences, we might be able to design vaccines based on weakened viruses and deploy them in record time.

Skiena's recent work started with the observation that not only are some individual codons unpopular, but some rarely occur next to one another. That is, there are unpopular *pairs* of neighboring codons. Skiena and his team are working on theories to evaluate why this is so. The problem lends itself to a genetic algorithm/mutational approach. Random codon sequences could be tried, mutated, and evaluated. The approach demonstrates a virtuous circle between computer science and biology: a biologically inspired algorithm aiding a computer scientist to solve a biological problem.

This is beautiful work with far-reaching effects. Skiena and his colleagues can design a viable virus from scratch. Nevertheless, for all the intellectual pleasure in understanding this work, we return to its scary aspects. What's to stop someone from designing a supervirulent virus by using popular codons or popular codon pairs? Skiena is skeptical. He believes putting in more popular codons or codon pairs would not create a fitter virus. If that were the case, the wild virus would have already selected that sequence for its genome. In fact, says Skiena, improving translation might make a strain that is too strong for its own good. "When a virus translates too well, it kills the host cell so quickly that it doesn't reproduce at all. The virus needs the host organism to reproduce. It's a complicated world out there, but we consider it very unlikely that anyone could juice up a virus using our techniques," he says. We hope he's right.

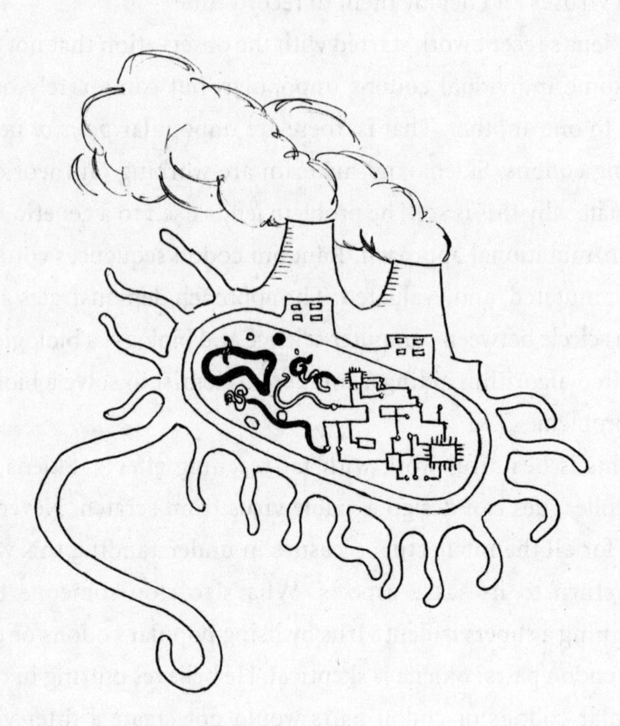

■ ■ ■ ■

The precision and reliability of embryogenesis in the face of constantly dying cells, their replacements, and changes in the environment—is enough to make any engineer green with envy.

— *Gerald Sussman*

Chapter 9

GERALD SUSSMAN

. . . .

Building a Billion Biocomputers

PHYSICISTS LEARN MORE THAN A SET OF PRINCIPLES ABOUT HOW the world works. They learn attitude—the confidence that they can figure out everything and build anything. After all, when your day job is to study the limits of physical laws, you may also want to challenge the limits of everyday life.

During the course of his career at MIT, Gerry Sussman has shown a consistent disregard for disciplinary boundaries. Though not a physicist by training, he wrote an advanced mechanics book because he disliked the existing mathematical notation. He created his own hardware design tools because he found the existing ones too inexpressive. With the help of some astronomers at Caltech, he built a "digital orrery" to solve an open problem from Isaac Newton's time about the perturbations of planetary orbits. And now, because he believes current electronic design is overly complex and expensive, he wants to build a billion-processor machine from bacteria. The concept takes its inspiration from embryogenesis, the method by which cells subdivide and create life. Sussman's attitude has always been the same—it's worth a try—and he brings to the effort an overriding optimism and sense of humor. He is also one of the founders of the Nerd Pride movement.

The son of a pharmacist, Sussman comes from Brooklyn, New York. He attended Erasmus Hall High School in Flatbush, the well-known spawning ground of celebrities in a variety of disciplines. The chess champion Bobby Fischer and Moe Howard of the Three Stooges are among its famous dropouts. Famous graduates include Barbra Streisand, Neil Diamond, and Sussman himself, who graduated in 1964. In high school, Sussman was a devoted tinkerer, and he loved to take apart radios and build new devices. He still tinkers. About twenty-five years ago, he took up watchmaking. "People younger than thirty don't have enough patience. People over sixty don't have good enough eyes," he quips.

As a teenager he also read everything about technology. In 1956, John McCarthy and Claude Shannon wrote *Automata Studies*, a seminal book on Turing machines and cellular automata, the foundations of modern computers. Inspired by that book, Sussman decided to go to MIT. At MIT he earned a BS and a PhD in mathematics (1968 and 1973). Although he studied math, his life revolved around computing, an unusual preoccupation in the computer-poor 1960s.

At the beginning of his freshman year, an upperclassman told him about Technology Square, where there were some computers that no one used, including a DEC PDP6. The DEC PDP6 was one of Digital Equipment's first large computers, a limited edition built for universities. Sussman was told, "You can go over and play with them." That was all he needed to know. Sussman became a regular at the technology center. "One day, this bald-headed fellow showed up," recalls Sussman. "I was sure he was going to throw me out because he obviously was the boss and thirty-five years old or something—an old man." The "old man" turned out to be Marvin Minsky, who had pioneered the field of neural networks. Instead of throwing Sussman out, Minsky asked Sussman if he would like to be paid for his programming efforts. Astonished, Sussman eagerly said yes.

Minsky asked why Sussman had made his tic-tac-toe program random. Sussman said the reason was that he didn't want it to have any preconceived misconceptions. Minsky replied, "Oh, it has them. You just don't know what they are." Sussman was impressed: "That was the smartest thing I had ever heard anybody say." He says he "became attached to this fellow" and worked for Minsky during the rest of his student years at MIT.

In 1951, Minsky had founded the field of neural networks with a randomly wired neural network machine. The idea was to create a network that emulated biological neural networks and changed connections in response to stimuli. A few years later, along with John McCarthy, Minsky founded the field of artificial intelligence. Following in their footsteps, Sussman's doctoral thesis would focus on machine learning.

Doctorate in hand, Sussman was still not certain what he wanted to do next. He enjoyed working for Minsky, but he was unsure what path his career should take. As he tells it, "Then one day some people at MIT said, 'We hear you're a good guy; maybe you want to be a professor or something.'" Sussman asked Minsky's advice about the offer. According to Sussman, Minsky told him, "It's a bad idea—professors have a lot of responsibility but no power." Sussman's wife, however, urged him to try it.

Sussman became an MIT faculty member in 1973, and he has been there ever since. His office is in the Computer Science and Artificial Intelligence Lab (CSAIL). It is the largest interdepartmental lab at MIT, with over eight hundred members. CSAIL is currently housed in a controversial Frank Gehry–designed building called the Stata Center (or the Geek Palace), which opened in 2004. With its irregular walls at different angles and its metallic sheen, the Stata Center looks as if it might take off, implode, or spin around itself.

One day in 1975, Sussman attended a lecture by Carl Hewitt, an MIT faculty colleague. Sussman sat next to a Harvard undergraduate named Guy Steele. Hewitt's lecture was provocative but difficult to understand. Sussman and Steele decided to discuss it further and ended up programming all night in an attempt to implement Hewitt's ideas. They soon realized they had re-implemented the lambda calculus (see the box "Lambda Calculus: A Quick History").

Sussman and Steele had developed a minimalist language called Scheme. Its combined simplicity and power make it a popular language for teaching programming. Sussman has argued convincingly that by programming in Scheme, students can learn advanced classical mechanics, which studies the motion of gases, solids, and planets, first described by Newton. He thinks programming in Scheme is far more effective than asking students to derive the underlying mathematical formulas, since the traditional mathematical notation can be ambiguous.

After taking a class from Lynn Conway, who worked on VLSI (*very-large-scale integration* of lots of circuits on a chip), Steele also discovered that Scheme could be used to design hardware that could execute Scheme. In 1979, working with Jack Hollo-

Lambda Calculus: A Quick History

Lambda calculus, a mathematical notation originally invented by Alonzo Church and Stephen Kleene in the 1930s, was the basis for the pioneering language LISP (short for ***list processing***) invented by John McCarthy in 1958.* Besides being heavily used in artificial intelligence, LISP has influenced other programming languages, largely because of its notion of ***recursion***. McCarthy invented LISP because he wanted to differentiate algebraic expressions. If you remember your high school calculus, taking the derivative of an expression sometimes involves taking the derivative of subexpressions, which is where recursion comes into the picture.

To understand recursion, consider the following definition of an ancestor: someone is my ancestor if he is my parent or an ancestor of my parent. The definition seems circular because ***ancestor*** is both the term defined and part of the definition. But a computer can turn this definition into a completely effective procedure. Given parent-child information and me as the starting point, my ancestors are my parents and the ancestors of my parents. The ancestors of my parents are their parents

(continued)

■ ■ ■ ■

* We include a biography of McCarthy and discuss his invention of LISP in our previous book, *Out of Their Minds: The Lives and Discoveries of 15 Great Computer Scientists.*

(continued)

and the ancestors of their parents. Just continue until there is no more parent information.

The lambda (λ) in lambda calculus enables a programmer to define a function without naming it. For example, as we learned in high school, it is possible to define $f(x) = x + 3$. That means the function named f takes an argument x and returns three more. So, for example, $f(5) = 8$. In the lambda calculus, one can say $\lambda x.x + 3$. So, $(\lambda x.x + 3)\ 5 = 8$. It is convenient to have no names for functions, especially when a program itself is going to generate them. Sussman and Steele explained how to program with lambda expressions and how to execute them efficiently.

way and Alan Bell, Sussman and Steele built a computer to run Scheme directly. The notion of a language used to create a chip to run that language had echoes elsewhere on the MIT campus. By 1980, MIT tour guides could show visitors robots that built copies of themselves. Humans provided the parts (not human parts), but robots did the rest.

Some technologists viewed all this as just the beginning. In a famous 1994 article in *Scientific American*, Marvin Minsky wrote that because of the limited life span of our biological components (livers, hearts, and even brains), robots or robotically enhanced humans should inherit the Earth. Sussman disagreed. Instead of working on biological organisms with silicon-based computational enhancements, Sussman envisioned computation based on biological parts. Biological parts are messy and hard to control, but they have one great advantage over the clean room and complex machinery necessary to fabricate silicon chips: they're basically free.

"Suppose you had computers that were the size of grains of sand and you bought them by the bucket and mixed them into concrete; you could have concrete by the megaflop. Suppose you could figure out some way to power them. Could you get the program to do anything interesting?" Sussman wondered.

Sussman thought that this multitude of entities wouldn't be in any particular local pattern, but globally in three-dimensional space they would communicate with their nearby neighbors via *amorphous computing.* In 1994, one of Sussman's graduate students, Andy Berlin, wrote a doctoral thesis about preventing a steel beam from buckling by putting sensors along the beam. If the sensors detected a buckling, they would control a mechanical device (something like "muscle wire" that tightens in response to a current) that would prevent the buckling. With a little power, Berlin achieved a major improvement in effective strength.

A competing school of thought founded by Sussman's friends Edward Fredkin, Norman Margolis, and Tommaso Toffoli had built and experimented with *cellular automata.* The automata themselves were invented by the mathematician Stanislaw Ulam and later simplified to a two-dimensional checkerboard game by John Conway in the 1970s, called the Game of Life. Imagine a very large, infinite checkerboard. At time t, the automaton assigns a value to each checkerboard cell that depends on its own value and its neighbors' values at time $t - 1$. Conway offered one rule for assignment. Fredkin, Margolis, and Toffoli offered many more and explored the applications of those rules to physics. More recently, Stephen Wolfram vastly extended the set of rules in order to discuss many branches of science in his book *A New Kind of Science.*

Sussman disagrees with this approach on three grounds. First, a cellular automaton grid is too regular, much more so than any biological system. Second, the cellular automaton assumes synchrony—all checkerboard squares update at the same time—which doesn't happen with real biological cells. Third, there is no provision for failures. In living organisms, cells die, mutate, and deform.

Sussman marvels at the ability of real biological systems to achieve near perfection in spite of their messiness. He cites ribosomes, which build proteins. "Ribosomes are humongous molecular machines—many molecules put together in a complicated way," he notes. Functionally, each ribosome is perfect and exactly like another. Sussman believes science should figure out how to harness ribosomes to make other things. Sussman and his graduate students—Daniel Coore, Radhika Nagpal, and Ron Weiss, as well as former student and long-time colleague Tom Knight—have developed programming languages to model and eventually control cell growth.

Sussman is intrigued by the long-term goal of improving on nature: "Wood is this marvelous stuff, but it is imperfect. It has knots and things like that. Well gee, wouldn't it be nice to make perfect wood? The cellulose in wood is glued together by lignin. Wouldn't it be interesting if we could make a particular bacterial species that would export the appropriate murein, a starch-like molecule, and put it together to make boards of wood?" Lignin is what makes cell walls firm, so more lignin means more durable wood. High-lignin wood with all the molecules lined up would be strong and light, making new wood structures possible. Sussman admits that his group is far from reaching such an outcome. "I'm just imagining thirty years in the future as usual," he says.

Achieving such a dream won't be easy. To understand the challenges, consider what an engineer wants from a material: control and predictability. Electronic circuits offer both. The materials stay where they are placed and have well-known physical characteristics that can be modeled mathematically. By contrast, biological cells are squishy and squirmy. And they die. Each is different from the other, making it seem impossible to do anything with such unreliable stuff. But embryogenesis—the process by which cells divide, replicate, and differentiate—proves that harnessing cells could work. The cells themselves aren't perfect, but groups of them form nose, skin, and heart cells with great precision. The question is how to make this phenomenon an engineering discipline.

Sussman's colleague Tom Knight has invented the notion of *biobricks*, building blocks for cellular computing. His inspiration came from the electronics industry. Capacitors, resistors, and transistors have to be made from iron ore and little pieces of gold and other special alloys. The industry has taken disparate raw materials and created interchangeable components. Knight has done something similar at the genetic level (see the box "Biobricks 101"). "Biobricks technology is about how to make standardized parts that could be hooked together in interesting ways. It's just the very beginning," says Sussman.

Suppose it's possible to create standardized biological parts. These could be tiny parts, a few micrometers (millionths of a meter) in diameter. They could be stacked and fed and maybe compute. There would be failures, of course, but Sussman believes that biobricks may have better fault tolerance properties than electronics have. "Biology has this wonderful property that if the situation changes, the animal or plant evolves. There's a built-in evolutionary potential," says Sussman. "You have to make something so that if it breaks, it grows around the failure. You don't want it to change when you're not looking either. You want adaptability, but with control." If such robustness could be wrested from the biological material, computing would become free, or nearly so. Because bacteria have mass and can potentially move, it is possible to do engineering as well as computing. Says Sussman, "I want to use bacteria to make an engineering infrastructure. I want them to be the fabrication devices—I want them to be the machine shop. And I want them to manufacture the molecular structures for the next generation."

Some might be concerned that we could end up with some kind of gray goo that consumes everything in its path. Sussman doesn't worry about such things: "Usually the arguments against technology are that we don't know everything. We can't solve the problem—there are things we don't know everything about." Sussman believes that not doing things is worse: "In general, nature proceeds as a car with no driver. Every so often, as in the Permian extinction, it runs into a tree. Unlike all previous species, we can grab the steering wheel. Now it may be true that we don't know how to drive, but we can learn. And if we don't try, we're going to hit the tree too, just like everything else has. Nothing has lasted more than a few million years."

Biobricks 101

The goal of the biobrick projects is to create a repository of circuit building blocks that can be hooked together. The component library should have an accompanying software tool that enables a designer to formulate a biological circuit on a computer and experiment with its behavior before committing to the biological substrate. The components will be placed inside cells that provide the machinery to keep energy flowing. The biobricks take the form of artificial gene-promoter pairs. For now, they are designed not to interfere with the cell's normal behavior (don't bite the hand that feeds you) nor with one another. The designers have made their first biobricks digital, to be sure to overcome intracellular noise.

To see how this is done, you need to know some molecular biology. DNA is composed of ***genes***, which produce proteins. Right next to each gene ("upstream" of it because DNA has a directionality) is something called a ***promoter***. For a gene to produce a protein, the promoter upstream of the gene must be bound by a number of inducer proteins but not by any repressor proteins. Certain promoters will cause their downstream gene to ***express*** (make RNA and then a protein) only if no repressor proteins attach to them. They don't need inducers—just the absence of repressors. We'll call these promoters "default-on." The most basic biobrick is an ***inverter*** made of a default-on promoter and an attached gene. The inverter will cause the gene to

(continued)

(continued)

express its protein only if the repressor is absent. That is, the inverter circuit is on when the repressor is off, and vice versa.

The next piece of molecular biology you need to know is that any gene-promoter pair can be synthesized. That is, any gene can be attached to any promoter. Suppose promoter **Px** is default-on but will be repressed by protein X. Suppose further that gene **Gx**, when expressed, produces protein X. Now, you could attach **Px** to **Gx** and get a negative feedback circuit. When X is present, **Px** is repressed, thus stopping **Gx** from being expressed. As a result, X gradually disappears, causing **Px** to induce **Gx** to produce more X. The effect is periodic expression.

Both X and Y are present, so Z is repressed

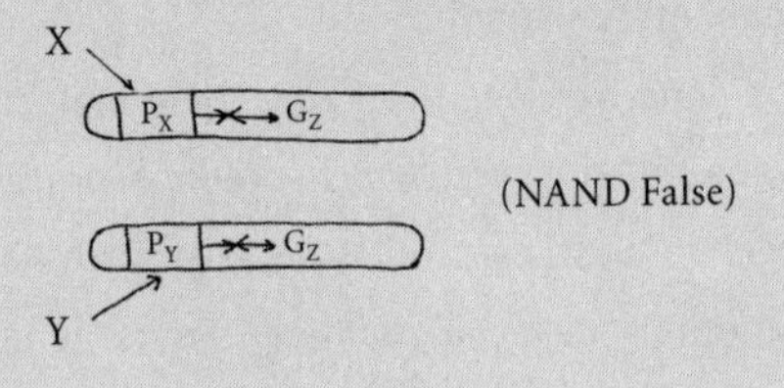

Either X or Y is present, but not both

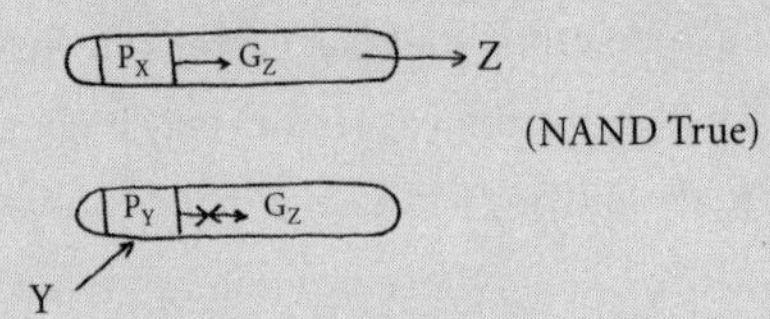

If proteins X and Y are both present, then protein Z will not be produced. If either one is absent, however, then one of these two promoter-gene combinations will fire and Z will be produced.

Other circuits are possible. For example, you could also attach ***Px*** to ***Gy***, leading to that workhorse of digital electronics, the ***NAND gate***. Any logical function involving AND, OR, or NOT can be produced by a circuit of NAND gates. In its two-input configuration, the NAND gate will produce an output of 1 as long as at least one of its inputs is 0. The NAND gate can be formed first, by the design of two promoter-gene combinations: ***Px-Gz*** and ***Py-Gz***. When both proteins X and Y are present, **Gz** will not be expressed, so there will be no Z protein. In every other case (if either or both of X and Y are absent—0), Z will be produced.

The inverter and the NAND are intracellular circuits. External signals (say, from humans) can reach these components, and cells are able to communicate with one another. Certain inducer molecules will diffuse through intercellular space and will enter through the cell walls as a result of concentration differences. The inducers modify the action of certain repressor molecules to prevent them from interfering with default-on promoters. The net effect is to turn on a switch.

The work is still in its initial stages, but the components are basically all there: circuit elements, an energy source, and communication. The biology is still messy, and it may well be necessary to interact with the cell some more if the goal is to control it. But this is definitely a good beginning.

■ ■ ■ ■

I don't want to program things that are mysterious. I want to program things that relate local to global, but in such a way that you can ultimately understand what's happening.

— *Radhika Nagpal*

Chapter 10

RADHIKA NAGPAL

■ ■ ■ ■

From Local to Global

THE HALLMARK OF MODERN DIGITAL COMPUTERS IS THAT THE smallest piece of information—the *bit*—is either 1 or 0. Another significant feature is mentioned less often: the "correctness" assumption. A software programmer assumes that the hardware will act correctly. By using detection circuitry and building in redundancy, hardware designers are able to compensate for failures and hide them from programmers. Programmers then need to think only about the correctness of their software, under the assumption that the hardware will work properly.

Now imagine that you are computing with thousands or perhaps billions of entities—biological cells or even electromechanical ones—embedded as sensors in a bridge or in the human body. Individual cells may fail at any time without warning. How can you do anything with a teeming crowd of unreliable components, especially components that can communicate with only a few neighbors? How would you even think about programming them? Those are the questions that Radhika Nagpal at Harvard's School of Engineering and Applied Sciences has set out to address.

Nagpal was born in 1972 in Atlanta, where her father was working on a PhD in mechanical engineering at Georgia Tech. When she was eight, the family returned to Amritsar, a city in northern India. Nagpal enjoyed school, especially art, mathematics, and biology. When she was in her junior year of high school, her father bought an early Intel-based personal computer to manage the accounting for his factory. He wrote a program in BASIC to keep the books. His daughter's job was data entry and finding bugs. One day, as she was entering data, the power went out and she lost everything. Power outages were frequent in Amritsar and Nagpal knew this would happen again, so she wrote a program to save the files. Then she taught herself how to write screen savers. She was hooked.

In India in the mid-1980s, a woman who did well in science was expected to become a doctor. Nagpal felt pressure to follow that path, but she decided she was not well suited to medicine. "The idea that people would come to me with lots of their problems wasn't at all appealing. I have enough problems of my own," says Nagpal.

As Nagpal was finishing up high school, Amritsar became a target of terrorist activity. In June of 1984, the Golden Temple, the Sikh holy site, had been attacked by the Indian army. In retaliation, Prime Minister Indira Gandhi was assassinated by her Sikh guards in October of that year. Conditions were getting worse. The government imposed a curfew, and if you left your house during the curfew time, says Nagpal, "all bets were off. It was just a really, really depressing atmosphere." (Despite the trauma of these events, Nagpal has retained a strong connection to Indian Sikh culture. She is active in Bhangra groups, the folk dance of the area, and contributes artwork to Chowk, a website for expats interested in retaining their heritage.)

When Nagpal thought about going to university, her main concern was getting away from the oppressive atmosphere in India. Her excellent test scores and US citizenship made going to school in the United States a possibility. When she was accepted at MIT in 1990, she decided to go there to pursue engineering.

One of the first classes Nagpal took as a freshman was MIT's introduction to computer science. The class used the textbook *Structure and Interpretation of Programming Languages* by Hal Abelson, Gerry Sussman, and Julie Sussman (Gerry's wife). Sometimes called the "wizard book," for the drawing of a wizard on its purple cover, the text has been the bible of MIT's introductory course since its publication in 1985. "It's a beautiful book—there are so many deep ideas in it. It was my first exposure to what computers could do and what you could do with computers," remembers Nagpal.

Nagpal plunged into computer science full force. She loved programming and entered every competition that was offered. She also had to take required physics and math classes, but having already taken those in India, she was bored. "Computer science was a savior for me," she says. The MIT approach to computer science requires students to learn by building. "You'd start very slowly: How does a circuit create memory? How does a circuit add? How does a circuit create a clock? You'd start from those basic elements and then all of a sudden you had built a computer," Nagpal recalls. She was fascinated that from such simple elements you could build something complex, with each layer building on something beneath it, and then from those modules you could create even more.

In those years, MIT had a program that enabled students to complete a master's degree with partner companies. Nagpal joined the Bell Labs program at a time when Bell Labs was still thriving. She joined a project led by Rae McLellan and Alan Berenbaum to build low-power, low-speed processors for early handheld computers. These included Apple's Newton and Eo's Personal Communicator. "Being a student in that kind of environment was exciting. They were not used to having young people. So when you became an intern there, all these famous people talked to you, and you heard their experiences in the thirty years that they had been working in computer science," says Nagpal.

Nagpal looked at Bell Labs and saw people enjoying life, inventing new things, and having intellectual hobbies. She returned to MIT in 1995, determined to get a PhD. She thought she would work on networking but soon decided it didn't allow enough freedom, because the Internet paradigm was so deeply ingrained. "I felt like I wanted a place where there was more freedom to explore ideas. I was attracted to having a totally new area where you get to invent," says Nagpal.

Daniel Coore, one of Nagpal's close friends from undergraduate days, was a student of Hal Abelson and Gerry Sussman, the founders of the "amorphous computing" group at CSAIL. Coore encouraged Nagpal to join them. By 1996, the group had just finished its first white paper on amorphous computing. Coore showed the white paper to Nagpal. "It was just out there," she says. She felt that they could invent the future with myriad little devices. It was parallel computing without the constraints of topology, but with sensors and actuators that could really interact with the physical world. Nagpal went to speak to Abelson and Sussman. She recalls their different styles. "Gerry would tell me the vision in everything, and then Hal would give me something concrete to chew on," says Nagpal. The team was trying to design a way to get millions of little simple devices to do something useful. The devices would work at approximately, but not exactly, the same speed. All communication would be to nearest neighbors, and devices would fail frequently.

In embryogenesis, creating an embryo is possible even if some cells fail. It is a complex process, yet somehow everything turns out all right in the end. It's a good thing that embryogenesis is so robust; otherwise, no species would propagate properly. Robustness was also an issue for pattern formation in such things as the stripes of a zebra. The question was how tiny failure-prone cells that operate at different rates (in computer lingo, "different clocks") multiply to make a stripe.

As it turns out, biology is economical. It uses a few mechanisms across many species. One mechanism involves a chemical gradient. A cell can "tell" how far it is from the source of the chemical by how much of a signaling chemical it receives. Chemical gradients require only local communication—chemicals diffuse from one cell to the next. Diffusion, in turn, transforms one of the chief challenges of amorphous computing to an advantage. The teeming millions of cells may all have unique shapes, but there are so many of them that their average properties are very uniform.

All cells that are equidistant from the source of the chemicals will receive nearly the same amount of chemical, even if some cells along the way have failed, and even if some cells are slightly slower than others. This observation leads to the possibility of programming macroscopic properties on top of asynchronous microscopic elements. For example, straight stripes emerge from tiny cells activated at different times.

An article in *Scientific American* by Christiane Nüsslein-Volhard, a Nobel Prize winner in biology, suggested a mechanism. Nüsslein-Volhard showed how simple chemical-gradient ideas are responsible for the exquisite patterning in the early development of the fruit fly. The patterning process itself was especially interesting: it applied the same idea over and over to create complex patterns. Nagpal then read works by Peter Lawrence and Lewis Wolpert on molecular and developmental biology. She was fascinated by how simple ideas in embryos can lead to complex global behaviors—from the ability to scale to various sizes, to the ability to regenerate parts.

To see how this works, imagine that you want to construct a black stripe in the middle of two points: A and B (the perpendicular bisector of the segment AB). There is a simple recipe to do this: Cause A to emit a chemical C_A and B to emit the same amount of a chemical C_B at about the same time. If an intermediate cell receives nearly the same amount of C_A and C_B, then color that cell black. The result will be a black stripe close to the perpendicular bisector of the segment AB. If the original amounts are different, then the stripe will be nearer to the source that emitted less chemical.

Nagpal credits Daniel Coore's "Growing Point Language" for showing the way to program an amorphous blob of unreliable cells, and calls it an "eye opener." He demonstrated how to use

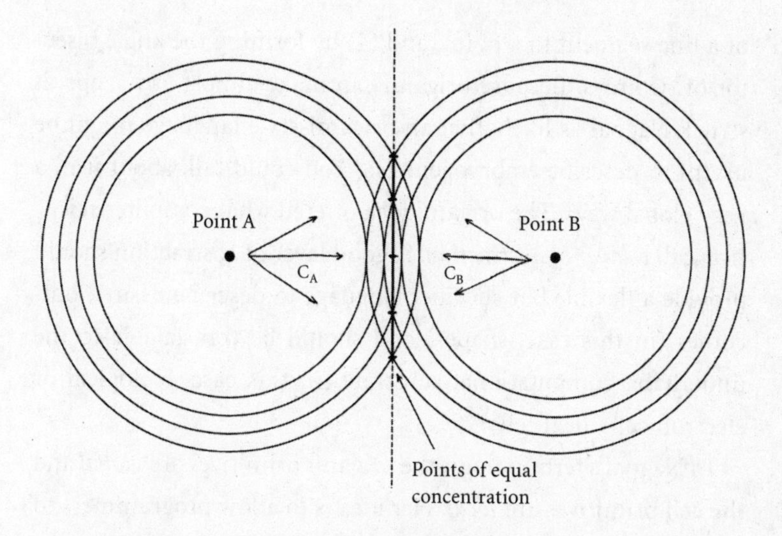

Each point in the perpendicular bisector of the line segment connecting points A and B will receive equal amounts of chemicals C_A and C_B. Points at the midpoint will receive the most of both chemicals.

simple gradients at different points in a computation to construct all kinds of patterns. Constructing patterns is a challenge, but for engineering a pattern must be turned into a shape. That's what an embryo does. "It's not just about pattern. Pattern is just the start of the process. Pattern can then direct change," Nagpal explains.

Actuators allow the possibility of forming folds—a kind of biological origami. The Japanese mathematician Humiaki Huzita proved that any origami shape could be created in four ways: (1) by folding a line segment between two points, (2) by folding from a point to a point (thus obtaining a crease at the perpendicular bisector of the two points), (3) by folding the endpoints of a line segment to a point, and (4) by forming the angle bisector of two rays. Because origami can make almost any shape, it struck Nagpal as likely that an origami-style language might be useful to describe embryogenesis: "You could talk about it in a very global way." The origami idea offered what computer scientists call a *layer of abstraction.* A good layer of abstraction should provide a flexible but succinct language to describe desired outcomes (in this case, shapes) and should be translatable to the underlying computational substrate (in this case, biological or electromechanical cells).

In Nagpal's terminology, the origami primitives are *global* and the cell primitives are *local.* Her idea is to allow programmers to write at the global level, and then a translator will convert the code to execute at the local level. Rather than trying to map a desired goal directly to the behavior of individual agents, the problem is broken down into two pieces: (1) achieving the goal globally and (2) mapping the construction steps to local rules.

Let's see what individual cells might do to fold a rectangle to a right triangle. Assume that cells can move, that chemical C_A is emitted from the lower-right diagonal point A, and that the same amount of C_B is emitted from the upper-left diagonal point B (see figure). Then each cell receiving more C_A than C_B can move toward the source of C_B (by moving toward its neighbors that have received more C_B). The cell can keep moving until the amount of C_A it received initially equals the amount of C_B it receives at its final destination. As a result, all cells will move from the A side of the perpendicular bisector to the B side and be evenly spread.

In her Origami Shape Language (OSL), Nagpal chose a different approach. Her assumption was that no cells can move;

instead they can cooperate to deform the sheet on which they lie. So if cells along a line cooperate and contract, they can cause the whole sheet to fold, in turn causing faraway cells to get closer.

Using moving cells or moving sheets, it is possible to create macroscopic moving sculptures with microscopic computing ele-

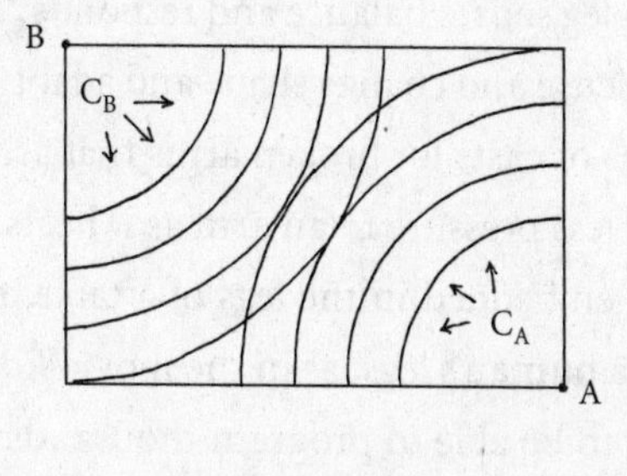

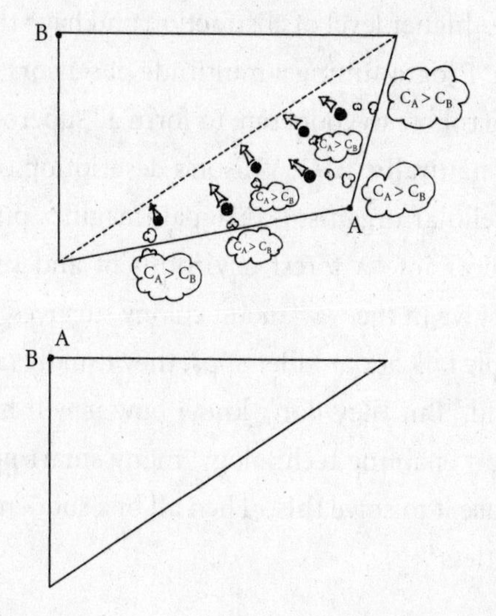

Cells in the lower right corner of the rectangle receive more of chemical C_A than C_B. That difference is a signal for the cells to move toward a greater concentration of C_B, causing a right triangle to be formed.

ments. This ability offers the potential of creating programmable materials, but Nagpal wants to go further. Biological cells sense the environment. That capability could help in a programming setting.

Nagpal and her group have designed a modular table whose legs lengthen or contract on uneven ground to keep the top of the table level—each leg senses balance and responds. The idea of this smart table is to sense and change shape and adapt to the environment. She dreams of casts for broken arms that are smart enough to apply more or less pressure to an arm as it heals. If sensors and actuators can be embedded in the legs of a table, they could also be embedded in a human's legs, as in the movie *RoboCop*.

Nagpal wants to be able to program thousands of sensors and actuators at a higher level of abstraction and have them do something useful. Programming a multitude of sensors and actuators could enable robots to cooperate to form a "superorganism"—in the spirit of naturalist E. O. Wilson's description of ant colonies and multi-cellular organisms. Nagpal imagines putting a group of small robots into a forest environment and seeing whether they can survive in the way an ant colony survives. She observes, "When people talk about killer apps, they usually talk about 'saving the world.' But they don't know how it will happen. When it becomes an enabling technology, many smart people can say, 'Hey, I can use it to solve this.' Then all of a sudden the possibilities are limitless."

Part III

PHYSICS AND SPEED

SINCE 1958, WHEN THE FIRST INTEGRATED CIRCUIT (OR *CHIP*) WAS introduced, the number of transistors (the electronic switches) that could be placed on chips has increased exponentially, doubling every two years. This trend is known in computer science as Moore's law, after Gordon E. Moore, who wrote a paper about the phenomenon in 1965. The speed of processors has also increased. Between the release of the Intel 4004 in 1971 and 2005, processors doubled in speed roughly every two years. That pleasant state of affairs resulted from miniaturization and faster clocks.

A computer clock is like a metronome that rises and falls at a certain rate. For example, a two-gigahertz clock rises and falls 2 billion times a second. Clock speeds rule performance for single processors: a processor performs one instruction between every two clock rises. Following an architecture first articulated by the mathematician John von Neumann, each instruction fetches data from computer memory (*RAM* or *cache* or a *register*), performs an operation, and returns the result to memory. Because hardware was improving so rapidly, programmers knew that their programs would run faster as soon as the next generation of processors arrived. That was the good news.

But there was bad news too. Increasing clock speeds also increased power consumption. A few years ago, the major chip manufacturers determined that increasing clock speed any further would require a power level that would endanger the integrity of the circuits. The manufacturers decided that the best way to take advantage of the ever-smaller transistors and wire widths was to put multiple processors on a chip without increasing clock speeds—creating so-called *multicore processors*. Getting the most from those machines entails writing and supporting *parallel programming*.

Monty Denneau has consistently designed computing machines that win speed competitions. His philosophy is to allow the problem to determine the geometry of his designs. He and his typically small teams have designed "petaflop" machines, nearly a million times faster than a current laptop. These machines solve complex problems in physics and have applications to code breaking. By optimizing design for a specific class of problems, Denneau is able achieve maximum performance. In his imaginative architecture, performance is enhanced because the data doesn't have to move too far. Imagine telling a customer that you can deliver a millionfold improvement to a product within two years, with only a handful of people on your team. This is what Denneau has achieved.

In many ways, **David Shaw** adheres to a similar philosophy, except that his focus is not on a class of problems, but on a single problem: the molecular dynamics of proteins. He begins with a model that pictures one or more molecules as a collection of atoms in three dimensions. He then computes how those atoms move in response to the forces they exert on one another. One requirement is to move the simulated atoms in small enough time steps to avoid approximation errors. A step this small lasts approximately a femtosecond (a thousand-trillionth of a second). Proteins require about a millisecond to do something interesting. There are a trillion femtoseconds in a millisecond. Each step requires computing the interaction among several thousand atoms. Shaw and his colleagues have designed a machine that simulates that step really, really fast.

Both Denneau and Shaw attempt to solve very challenging computational problems by designing digital electronics that conform to the physics of the problems. It's possible that digital electronics is not the best hardware medium, however. Proteins fold in a millisecond, but even Shaw's tour-de-force computer, Anton, takes days to simulate the fold. To avoid making mistakes in arithmetic, digital electronics forces the values entering memory to be either 0 or 1. But the underlying electronic elements—the transistors and resistors—are able to produce an infinite number of voltages. Some scientists have found a way to take advantage of that fact.

When trying to model problems in nature, **Jonathan Mills** uses continuous physics rather than discrete electronics. Continuous problems in nature include the movement of particles in a force field or weather patterns. Inspired by the work of the mathematician Lee Rubel, Mills has built a machine called the *extended analog computer.* This machine can "solve" the differential equations that characterize many problems in physics. The machine itself contains a material (typically some kind of foam) whose physical properties model a variety of differential equations. You can program Mills's machine to solve a physical problem by causing the material to mimic the problem and then measuring the voltage outputs of the material. Mills finds this approach far more natural than simulating the differential equations through lots of calculation. To date, his results are preliminary but persuasive.

Scott Aaronson pursues an entirely different alternative to the digital paradigm: computing at the quantum level. Although each bit in a digital computer can be in only one state (0 or 1), a quantum bit (*qubit*) state can be partly a 0 and partly a 1. This here-and-there property can be extended to many qubits, allowing a set of 20 qubits to encode (collectively) over a million 20-bit combinations. If a programmer could encode each possible solution to a difficult problem and pick out the combination representing the best solution, then a whole class of important but still unsolved problems (called *NP-complete*) could be solved efficiently. It's not at all clear that this is possible, but Aaronson is determined to find out. Our understanding of nature (including the possibility of time travel) may depend on it.

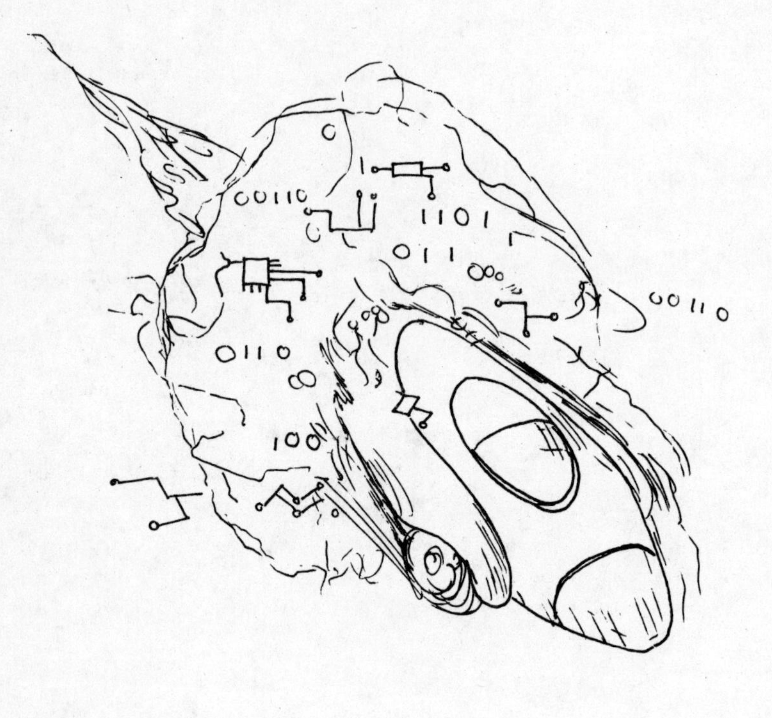

■ ■ ■ ■

It is sort of a God's-eye view where you see what every single processor would do on a single clock cycle and could change what that was. It was so perfectly designed for what it does. It's kind of like when someone designs a racing-car engine and gets it just right.

— *Monty Denneau*

Chapter 11

MONTY DENNEAU

. . . .

The Architect of Speed

COMPUTER ARCHITECTURE ATTRACTS AN ECLECTIC ASSORTMENT of people—mathematicians and engineers, even musicians and visual artists. Often they enter the field because they understand intuitively how to map a conceptual problem into an appropriate geometry. But the geometry is just the blueprint. Like a building architect, the computer architect must figure out how to deliver power, remove heat, and deal with failures—and do it all economically. Unlike a building architect, the computer architect must take risks on unproven technologies, yet still deliver a product that rarely fails.

Monty Denneau of IBM's Watson Research Center in Yorktown Heights, New York, is the architect of a recent *petaflop* machine—it does a million billion operations per second. He came to design from mathematical philosophy. How he got there, in retrospect, makes a lot of sense.

Born in 1948, Denneau was raised in Lincoln Park, New Jersey. His father started out as a car salesman but ended up in factory work. During World War II, Denneau's mother used an electromechanical calculator while working in an airplane design shop. When she couldn't get a similar job after the war, as those jobs reverted to men, she went to work in a factory, sometimes on the graveyard shift. (The kids learned to cook for themselves.) Both parents encouraged education. "My mother decided that I should be a doctor. She even made me take Latin because she thought doctors had to know Latin," Denneau recalls.

Denneau's educational path encountered bumps, mostly because he found it difficult to remain quiet in class. Even now, as a senior designer, he talks about his supercomputing work with an infectious enthusiasm. But as a child, this unbridled energy caused problems with teachers. Fortunately, he fell in love with mathematics. His Estonian math teacher sat him in the back of the class in his own little corner. While the rest of the class worked on something else, Denneau would work on his own projects. "She kind of rescued me. Otherwise I would have been doomed. I was just fooling around and getting in trouble," says Denneau. He also began tinkering with electronics, mostly ham radios.

Denneau ended up at MIT, but not to prepare for a medical career as his parents wished. At first he was drawn to philosophy—ontology and theology. He characterizes himself as a "dumb kid" who grew up late and was just doing things that were interesting to him without thinking of practical considerations. Finally, he majored in mathematics. To his regret, Denneau missed being mentored: "I really wasn't mature enough. I was thrown into a dorm with a bunch of other people, probably half of whom had Asperger's syndrome."

Denneau earned a master's in mathematical logic at Boston University and then continued at the University of Illinois. His

PhD advisor, Gaisi Takeuti, gave him a problem for his thesis. It had to do with infinite-dimensional *Hilbert spaces.* Hilbert spaces are important for quantum mechanics and signal processing. They extend the geometry of two and three dimensions to infinite dimensions. This particular aspect had not been worked on very much. It was unclear whether Denneau would be able to solve it. But one thing *was* clear: if he didn't solve it, he wouldn't get his PhD. "When I think back to that time," Denneau recalls, "I think, how could I have taken that risk? As it happened, I just kept plugging and plugging and plugging. I eventually solved it." That same combination of risk taking and hard work have informed the rest of his career.

Denneau thought computer science was for engineering types and didn't take his first programming course until graduate school. The course taught BASIC, a rudimentary programming language, and he liked it. He went on to a computer engineering class in designing circuitry. His path to becoming a computer architect really began when he entered a contest and designed the best circuit out of a hundred entries. Denneau was surprised at his victory, since he had no engineering background and still thought of himself as "a philosophy guy." He went on to build his own computer, even designing a graphics card. In parallel with his doctorate in math, he earned a master's in computer science in 1978.

Denneau claims to have "bluffed" his way into IBM. Bluff or not, he proved his worth with his very first project, the *wiring machine*, one of the first parallel processors, which was built with 64 microprocessors in the early 1980s.

Every digital chip includes transistors and resistors, as well as wires connecting them. If two wires touch, they create a short circuit. If the wires were placed on only one planar layer, the possible interconnections would be very limited. For this reason, the chips contain several layers (analogous to floors of a building), which allow for more complex wiring patterns and extremely challenging layout problems. IBM asked Denneau to build a machine to figure out the best wiring layout for IBM's high-end machines.

Denneau put together an 8-by-8 grid of Z80 processors—64 microprocessors in all. His idea was to make a correspondence between the processors and the chip to be wired: Each processor would be concerned with a small area of the chip. The processors would communicate about wires that had to cross the border between areas. In that way, the Z80 processors could work mostly independently and in parallel. Denneau learned an important lesson from this project: "the need to match the computing to the work that has to be done."

IBM at that time ruled the world with big, single-processor machines, and the idea of using two processors was a stretch. Using 64 was considered foolhardy. "Everybody said this was ridiculous. You can't possibly have 64 processors working on a problem and cooperating. Fortunately, we just went ahead and did it," Denneau recalls.

In linking 64 processors together, Denneau encountered two problems that plague any parallel-machine designer. First, he needed reliability—machines break all the time. Second, he had to program a gaggle of processors to do useful work side by side without chatting one another to death. By matching the computing to the work, he was able to ensure that each Z80 could do most of the necessary work on its own, requiring only occasional communication with the other Z80s.

The wiring machine showed the promise of a new paradigm:

parallel machines could work much faster than mainframes. Ironically, this paradigm almost led to IBM's downfall two decades later, when groups of microprocessors replaced much more expensive mainframes.

Denneau's next machine was called the Yorktown Simulation Engine (because IBM's Watson Research Center resides in Yorktown Heights). Building chips is a multi-million-dollar operation, and if you make a mistake you have to start all over. The Yorktown Simulation Engine was designed to simulate the logical operation of the chip before it was built. (There were other software simulators, but Denneau's was thousands of times faster. A huge switch interconnected the computer boards and enabled them to simulate the signals going from one circuit to another.) "You could really do this only at IBM, because it was risky to start with. The good part was that we got a thousand times return on our investment," says Denneau.

In general, there are two ways to get speed: good algorithms or good machines. Certain problems are clearly suited to a parallel machine. For example, consider a problem with many mutually exclusive scenarios, each of which must be tried. In that case, each processor in a parallel machine can try out a scenario without interfering with any other processor. When finished with one scenario, the processor can work on a new scenario. Finance companies, for example, use parallel processing to explore the implications of different interest rate scenarios on bond values.

Other types of problems require occasional interaction. Local government provides an everyday analogy. Most city-planning issues, such as land use, affect no other city. Other issues, such as transportation routes, can involve many cities in the county, state, or region. Although mayors can make most decisions on their own, they occasionally need to communicate with their colleagues elsewhere. Denneau's wiring and simulation problems enjoy this mostly local property. Other problems are truly global in nature and present the greatest engineering challenges.

In the mid-1980s, some physicists asked Denneau to design a machine to generate predictions using the theory of *quantum chromodynamics*, which describes the interaction of subatomic particles like gluons and quarks. It wasn't clear whether the theory worked, but it was easy to test theoretical predictions experimentally. The problem was to compute those predictions.

Arithmetic calculations depend on previous calculations: You have to wait for the results and fetch new data, which might require five clock cycles to get something done. "You start a calculation and then you're sitting there like an idiot," explains Denneau. Those delays, called *pipeline stalls*, prevent many machines from achieving full performance. Another problem is memory. A machine asks for something from memory but in the meantime doesn't have enough work to do—it's underutilized. A third problem is communication. The machine uses packets to communicate—units of binary data, 0s and 1s. It sends a packet request, and all the packets fight with one another to get through the network.

Overcoming these problems required what Denneau calls "compiling out the calculation." It helped that quantum chromodynamics involves the repetitive execution of many similar operations. *Compiling out* means taking advantage of the special properties of the computation to move data just where it's needed when it's needed. Denneau learned to fine-tune the machine to the problem until it purred like a racing-car engine. He was able to build an 11-gigaflop (11 billion instructions per second) machine called GF11—hundreds of times faster than commercial machines of the day.

In the 1990s, Denneau became intrigued with the problem of sending continuous, complete images over transmission lines. He wanted to find a way of reconstructing the image rather than sending compressed pixels. Doing this efficiently required fast execution of a computer graphics algorithm known as *ray tracing.* "The problem was that ray tracing ran terribly on all machines, in particular on our own machines," Denneau admits. The machines of the time were designed to bring big blocks of data into the *cache line*—a temporary memory storage area. Unless you could use all that data, you would get poor performance.

Ray tracing works backward and involves reversing the way light hits the eye. To determine the color and saturation of the multi-million-pixel panel representing the eye's light receptors, ray tracing imagines sending a light ray away from the pupil. That ray might hit something red and then stop, so the pixel should be red; but if the ray hits something reflective, it keeps going and collects more information. Data is collected from everywhere in the scene, but in an unpredictable way. So each ray must gather data from many places in memory. Denneau realized that ray tracing was not feasible on ordinary machines. Ray tracing had no locality.

Denneau started imagining what a ray-tracing machine would look like. He figured out that the machine needed "simple, little dumb processors" that would do the operation and then would sit there. It didn't matter if one processor was idle, because it occupied so little of the silicon. Denneau called these simple, dumb processors *thread units.* Each thread unit would be responsible for one ray. Unfortunately, as often happens when someone invents beautiful technology, there was no market. Nobody cared that teleconferencing pictures could become photorealistic. Fortunately, as also often happens, cool technology has a way of finding new applications, if you can seize opportunities.

One day, Denneau bumped into a colleague at the Xerox machine on his floor. His colleague described a problem called *protein folding* and said there was no computer remotely fast enough to solve it. He asked Denneau if his ray-tracing machine could do it. After tweaking his original design a bit, Denneau realized that his machine would work well for protein folding. This chance meeting occurred in the late 1990s. Denneau and his team put together a proposal that launched what became known as the Blue Gene project. "I'm proud of having invented the name," says Denneau. But the head honchos at IBM got worried. They had never built a chip with 160 processors (thread units) before. Most people thought that was an unreasonable thing to do. And it was not considered an official IBM architecture.

Then, "an unspecified government client" at an undisclosed location became interested. It turned out that Denneau's approach worked not only for ray tracing and protein folding, but for a lot of other important problems. Denneau never told us who his client was and we knew not to ask, but various code-breaking agencies have a seemingly inexhaustible appetite for high-performance computers. The National Security Agency, for example, has funded many of IBM's projects since the 1950s.

When it was actually built, the Blue Gene machine ended up using a more classical architecture of only two processors per chip. Even with classical architecture, it has achieved sustained 500-teraflop speeds (500 trillion numerical calculations per second) on applications like protein folding and quantum chemistry.

Government financing has enabled Denneau to build his thread unit machine, now called Cyclops. Denneau's view of dumb processors has radically simplified processor design. Building "smart" processors like the ones in your laptop involves guessing what the next instruction will be, providing for backup if that guess turns out to be incorrect, and constructing a very large set of operations for text and numerical processing. That paradigm requires hundreds of designers and lots of power. By contrast, Denneau was interested in lots of simple processors on a chip, with each one executing one instruction at a time. In fact, the processors are so simple that several of them share "floating point" hardware (to perform arithmetic operations on fractional data) and fast memory.

Denneau predicts that Cyclops will be one of the first petaflop machines. He put it together with a tiny group of people—"5% of the usual resources," he says. "I have remarkably talented friends and students," says Denneau with pride. "They're a unique group."

With so many processors, communication delays can dominate the time of a computation. Denneau chose a crossbar switch that resembles a large checkerboard pattern. Each processor X is associated with row X and column X. Processor X communicates with Y by sending a message along row X to column Y, where the message is forwarded to processor Y. This configuration provides a two-step connection between any pair of processors but requires an enormous number of wires (technically, the square of the number of processors). Denneau's colleague Gregory Chernasky ("one of the smartest guys I know," says Denneau) figured out how to route the wires on the chip, but that task added many months to the schedule.

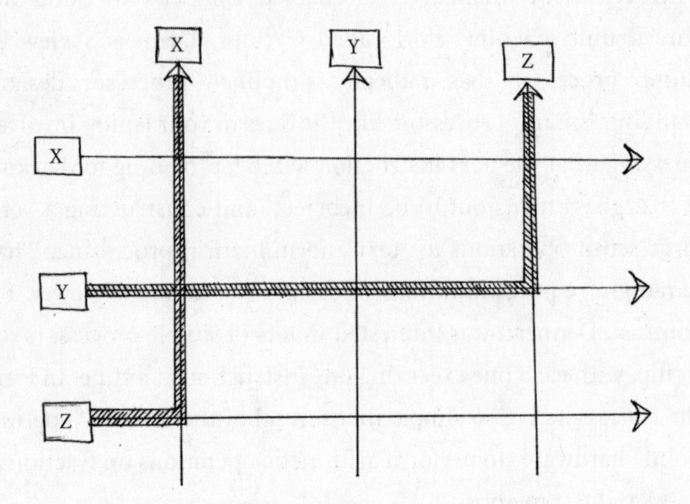

A message from row Y to column Z can proceed at the same time as a message from row Z to column X because the two messages share neither a row nor a column.

Fault tolerance proved to be an even bigger problem; and each year, as circuit elements decrease in size, it becomes more of a problem. Alpha particles and cosmic rays hitting smaller circuit elements can cause errors. Denneau therefore built in redundancy. Cyclops was going to have two chips side by side that would do the same calculation, but the client didn't want to pay for double the power—double the megawatts. Instead, Denneau and his team designed the machine so that many of the calculations would be self-checking. You do a rough calculation and if you don't get it, you do it again. In some cases, you run it twice. "That's how we're dealing with it now. That may or may not be a good thing for the future," Denneau admits.

Failures can occur even before the machine is turned on. Making the chips requires extremely clean room conditions because a single piece of dust can completely ruin a chip. The Cyclops chip is one of the largest that IBM has ever built. Large chips typically have low yields. That is, you might manufacture 100 chips but only 10 pass all tests (a 10% yield). The larger the chip, the more things can go wrong.

People once again were skeptical. Denneau reasoned, why do all 160 processors need to be good? What if only 158 were good? Replacing the chip is time-consuming and ultimately self-defeating. "The more you touch the machine, the more you break it," he says. So Cyclops just continues to run, circumventing failed processors. Denneau thinks his fault tolerance approach will catch on: "It's going to become universal because you're not going to be able to get enough yield on the large chips."

Denneau's customers are addicted to speed. His next machine may run in the scores-of-petaflop range. One current application supports data processing for large radio telescopes. Current successes don't necessarily imply future ones, however. In fact, Denneau says, "we can't make progress as fast as we've been making it." He cites limitations in the laws of physics. To reduce power, voltages delivered to chips have been decreasing over time. Reducing the voltage too much may make it possible for chips to mistakenly flip 0s to 1s, or 1s to 0s. Another problem is bandwidth between chips. Denneau is looking forward to direct optical connections (such as lasers) to the chip instead of wiring. Until then, Denneau says, "we're in trouble."

For single applications dedicated to one purpose, Denneau has figured out how to minimize communication, use self-checking computations to manage failures, and keep the power and cooling reasonable. "The other problem with all these things is the integration with software. It just stinks," he says. "They're programming the same way they did 30 years ago—in C and C++. The tools are still poor."

Denneau thinks that software inertia comes out of organizational inertia. It's hard to get people to use an innovative language if nobody else is using it. In fact, he thinks software development has regressed, at least for scientific calculations. He makes reference to APL, which was "a very cryptic language that looked like Phoenician. The whole screen would fill up with a matrix and solve problems." As you may recall from our discussion of Jake Loveless's work (see chapter 4), there is a small but very productive cadre of programmers using highly expressive array languages that derive many of their ideas from the APL language that Denneau mentions. With the ability to think of entire arrays of numbers instead of individual numbers, a programmer can express a problem much more succinctly. For example, the shortest Sudoku program in the language K requires fewer than 100 *characters.*

Brevity for the human programmer is only one of the benefits of array languages. Brevity also helps the machine compiler do its job. The compiler is the program that translates the language commands into codes that the machine can understand directly. When the programmer states, in a single operation, "Multiply this 1-million-element vector by that one," the compiler can spread the work to many parallel processors. Compilers have a much harder time trying to figure out how to distribute work when programmers use "scalar" languages in the C language family, in which they examine each element one at a time. The same holds for data manipulation operations such as, "Find all employees whose age lies between 40 and 60." The ability to perform a single operation on large amounts of data opens the door to vast performance improvements on parallel machines. Someday soon, every computing machine will be a parallel machine.

Linguists Edward Sapir and Benjamin Whorf famously asserted that language influences how humans think and what we can think about. Similarly, computer languages influence how and what the machine calculates. Software programmers for Denneau's massively parallel machines will need a language that makes thousands of processors purr.

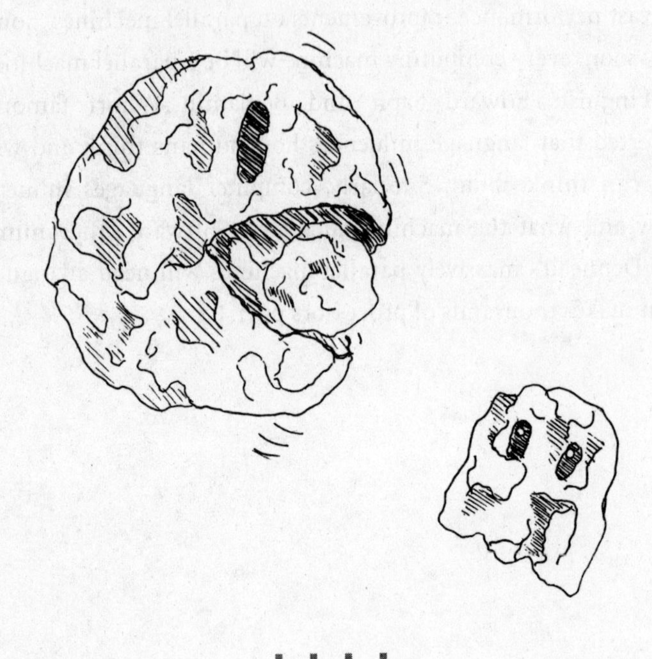

■ ■ ■ ■

The most important thing is the possibility of someday making a small contribution toward saving lives. Not a foregone conclusion that we will. We'll probably learn some scientifically interesting things, but nobody knows what the real impact will be on drug discovery.

— *David Shaw*

Chapter 12

DAVID SHAW

....

Anton and the Giant Femtoscope

"THE REASONABLE MAN ADAPTS HIMSELF TO THE WORLD: THE unreasonable one persists in trying to adapt the world to himself. Therefore, all progress depends on the unreasonable man." So said George Bernard Shaw. He could be describing David Shaw (no relation).

David Shaw is unreasonable in the best sense of the word. For starters, he has accomplished what many others only dream about. The investment firm D. E. Shaw & Co., which he founded in 1988, has made him fabulously wealthy and has completely changed the nature of mathematical financial trading. The company's philosophy has been to hire great people, give them outstanding technical support, and then ask them to exploit inefficiencies in financial markets.

As Shaw approached the age of fifty, a chance conversation with a friend some forty years his senior convinced him to apply his quantitative and computational talents to a realm completely outside the field of finance: understanding the atomic-level behavior of biological systems—a field known as *molecular dynamics.* His current goal is to speed up the solution of problems in this field by a factor of 1,000.

Born in Chicago in 1951, Shaw was raised in Los Angeles. Early on, he loved science. At age three, he decided he wanted to be a medical doctor. A bit later, "after several years of dismembering potted plants and canned sardines, I talked my parents into buying me a surgical scalpel and letting me dissect rats in our basement," he recalls. Shaw's father was a plasma physicist; his mother, an artist and academic with a PhD in art education. His parents divorced when he was about thirteen. Shaw's stepfather, who was a professor at UCLA's School of Management, provided him with the model of the life of a researcher.

Shaw completed his undergraduate work at the University of California at San Diego, where he studied mathematics and applied physics. He became interested in computational models of cognition after taking courses taught by David Rumelhart (a pioneer in the field of neural networks) and experimental psychologist Donald Norman. At their suggestion, Shaw pursued his doctoral studies at the artificial intelligence laboratory at Stanford. The Stanford AI Lab was a vibrant place involved in all aspects of the field, including robotics, computer vision, machine learning, speech recognition, and natural language understanding.

Midway through his graduate studies, Shaw took three years off to start a small software development and consulting company. He then returned to Stanford to finish his doctoral work under the supervision of co-advisors Gio Wiederhold and Terry Winograd. Shaw's dissertation proposed a way to speed up certain types of artificial intelligence and database management applications using a new type of massively parallel machine. The thesis was theoretical—he never actually built any hardware. At the time, Shaw knew almost nothing about the practical aspects of a computer system.

After finishing his PhD in 1980, Shaw joined the faculty at Columbia University, where he led a project to assemble a working prototype of the machine described in his thesis. The system was built with a custom integrated circuit, designed in the Shaw lab. The circuit included a number of small processors, each with its own memory. By integrating logic and memory on the same chip, the machine avoided the so-called *von Neumann bottleneck*—the time required to communicate data between processor and memory.

The von Neumann bottleneck limits the speed at which applications can be executed on a conventional computer system. Shaw's "Non-Von" machine was a highly parallel machine for knowledge-based systems. It was not intended to be a general-purpose or dedicated scientific supercomputer. At the time, Seymour Cray was designing those kinds of pioneering supercomputers. Instead, Non-Von was designed for complex pattern-matching problems that might come up in searches by "meaning" (as opposed to simple keywords).

After completing the Non-Von prototype, Shaw became interested in developing massively parallel machines aimed at a broader range of applications. Although the Non-Von project was very well funded, by academic standards, Shaw soon realized that the next machine would require a much larger budget than was likely to be provided by the government agencies that had so far supported his research. He thought it would be a good time to look for commercial funding. By the time he finished his business plan, however, he found that venture capitalists weren't interested in start-ups in parallel processing.

While pitching his machine, Shaw spoke to several organizations, including the investment banking firm Morgan Stanley. It turned out the firm had recently discovered certain anomalies in the stock markets and had developed a computer program that was automatically generating highly profitable stock trades using the bank's own capital. In order to fully exploit the potential of this new type of trading, the firm was looking for someone with Shaw's background to bring new technologies to bear on its ongoing research effort. Shaw had never thought about Wall Street. But he got excited about using technology in the world's financial casino. Another inducement was the salary that Morgan Stanley offered—six times what he was making as a Columbia professor. "I realized that accepting would also allow me to quickly save enough to comfortably support a family at some point," Shaw recalls.

In 1988, Shaw left Morgan Stanley to found D. E. Shaw & Co., which he envisioned as an academic-style research group focused on identifying investment opportunities throughout the world's financial markets. Several financial economists had by then laid the groundwork for quantitative trading, and a handful of investors were already experimenting with systematic strategies based on relatively simple formulas.

D. E. Shaw & Co. served as a prototype for a new generation of investment firms that would apply advanced quantitative and computational techniques to formulate large, intricately constructed portfolios. The goal was to achieve unusually high returns while systematically managing various forms of risk. Shaw was among the first of what are now referred to on Wall Street as *quants*, and he has a remarkable track record. "Computational finance was still pretty much virgin territory when the firm was first getting started, so there was a lot of low-hanging fruit," Shaw recalls.

As the firm grew to more than a thousand employees, with about $40 billion in invested capital, Shaw found himself spending more time on management and less time on technical work. "I missed doing real science," he remembers. "In an attempt to keep that part of my brain entertained, I started working on random applied mathematical problems on my own at night and in the shower."

Shaw's recreational activities became more focused when a friend, Columbia chemistry professor Richard Friesner, started giving him small computational problems to solve from his research on the structure, dynamics, and function of proteins. Earlier, Shaw had learned about the so-called protein-folding problem from Cy Levinthal, who had been chairman of the biological sciences department at Columbia when Shaw was on the computer science faculty. Levinthal knew Shaw was building massively parallel machines and wondered if one of them could simulate the folding of proteins. Though fascinated by the problem, Shaw didn't have time to pursue it while running D. E. Shaw.

For many years, Shaw ruminated over the possibility of going back to full-time research. In 2001, on a bus ride with Bill Golden to Shaw's fiftieth birthday party, Golden, then ninety-one, encouraged Shaw to follow his dream. As a young man, Golden had also made a fortune on Wall Street and had left to pursue his own dream of bringing scientific expertise to the highest levels of government. Golden helped instigate the founding of the US National Science Foundation and talked President Truman into creating the position of Science Advisor to the President. Holders of that position have guided successive administrations through the challenges posed by nuclear weapons, the space race, environmental and energy policy, and a host of other issues. Over a period of more than six decades, Golden was a mentor to many fledgling members of the American science and technology policy community, including Shaw, who had been appointed by President Clinton in 1994 to the President's Council of Advisors on Science and Technology.

"When Bill Golden said 'Just do it!' I took it seriously," Shaw says. Science was what Shaw enjoyed most and what he had been trained to do. He was now in the enviable position of being able to conduct self-funded research. Golden's advice, together with the model of Golden's own life trajectory, led Shaw to appoint himself chief scientist of D. E. Shaw and delegate everyday operations of the financial side of the firm.

Shaw thought protein dynamics problems provided a good fit for his interests and his background in developing novel machine architectures. He had a personal reason as well: his mother, father, and sister had all died of cancer. Although he had no training as a medical or pharmaceutical researcher, he wanted to design a computational tool that would someday be used to develop lifesaving drugs. "I had no illusions about being a general in the war against cancer," Shaw says, "but it felt good just being a soldier."

Molecular dynamics examines configurations of multi-atom structures, along with the atoms in their environment, and then determines how the configurations change over time in response to the forces among the atoms. Proteins typically have thousands of atoms. In addition, there are surrounding water molecules. There are no known "closed form" formulas to predict this—you can't just solve the problem by manipulating equations. You have to simulate.

To conduct a molecular dynamics simulation, you divide time into very short time steps, on the order of a femtosecond each. A femtosecond is 10^{-15} second, a million-billionth of a second. During each of these time steps, you calculate the forces among all of these atoms and then calculate where each atom would move over the next femtosecond as a result of these forces. You then repeat the process an enormous number of times.

In Shaw's case, *enormous* means as much as a trillion because most biologically interesting movement (for instance, the folding of a protein or the docking of one protein onto another) takes place on a timescale approaching a millisecond. Each instance of calculating forces, if done naïvely (using the first method of calculation that might spring to mind), would require looking at the effect of every atom on every other one. That process translates to hundreds of millions of interactions for pairs of protein-and-water molecules containing only tens of thousands of atoms. Simulating a femtosecond might require several seconds on a conventional processor. The trillion time steps could then take more than 100,000 years. Massive parallelism can help solve many problems, but this problem is fundamentally sequential: one femtosecond must be done before the next one. The challenge is to make the fundamental femtosecond-level simulation faster. This has been Shaw's focus.*

* An alternative approach to inferring the structure of a protein or a protein complex is to use the databases of known protein structures (determined by X-ray crystallography) to help predict unknown structures. Certain short amino acid sequences, for example, often fold in a similar way in many different proteins. A highly successful collection of software called Rosetta, originating in David Baker's lab at the University of Washington in Seattle, embodies this approach. Rosetta has successfully pre-

Fortunately, several methods can significantly reduce the amount of computation required for the atoms separated by a relatively large distance. Although the details differ, each of these approaches accurately approximates the effects of widely separated atoms. Even when such techniques are used, however, it's necessary to compute the forces of each atom with respect to all of its *nearby* neighbors. The question is how to do that fast.

If the information concerning each atom is localized to one processor in a parallel machine, we can imagine several approaches. The most natural approach would be to send atom A's information to the location of every neighbor B (see figure). In Shaw's terminology, this means sending A's information to "the territory" of every neighbor B. Symmetrically, A would receive, in its territory, the information concerning every neighbor B. But this scheme still requires a lot of data movement. Shaw's method uses the geometry of the problem to cause the information transfer from A to B to take place in "neutral" territory—neither at the A location nor at the B location (see the box "Computing in Neutral Territory").

Neutral territory methods can speed molecular dynamics calculations for conventional parallel clusters, but Shaw has gone beyond algorithm design to the construction of his own parallel machine, called Anton. (Anton is named after the seventeenth-century maker of microscopes Antoni van Leeuwenhoek.) Shaw's Anton machine contains 512 custom-designed chips, each containing, among other things, 32 specialized 28-stage arithmetic pipelines designed specifically for ultra-high-speed calculation of

---

dicted the folds of proteins in several competitions and enjoys growing use in the pharmaceutical industry. Ultimately, a hybrid strategy that learns from experience and that uses molecular dynamics may emerge.

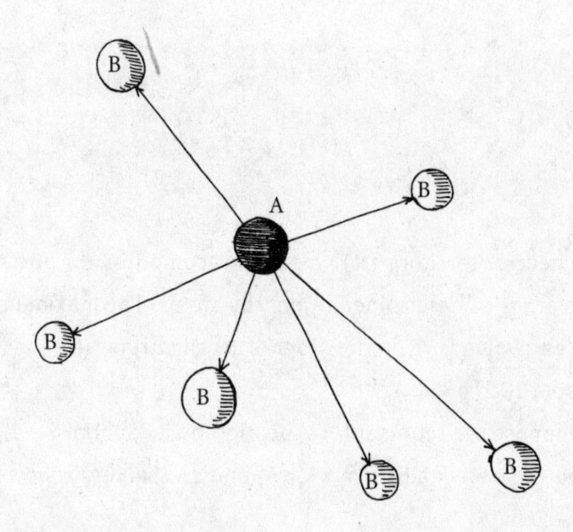

Each atom depends on forces from all other atoms. Naïvely, this means sending information from every atom to every other atom.

the pairwise particle interactions involved in molecular dynamics simulations. Each pipeline, in turn, contains a large number of tiny arithmetic units and is capable of producing results during each 800-megahertz clock cycle that would require approximately 50 arithmetic operations on a conventional processor.

Working together, the pipelines embedded in Anton's 512 chips are thus able to execute the most computing-intensive portion of a molecular dynamics simulation at an effective peak speed of more than 650 trillion operations per second. Additional computational power is provided by a number of more flexible on-chip processing elements. In its normal mode of operation, Anton performs these calculations without ever accessing any off-chip memory. The result of one calculation often flows directly into another processing element. Anton avoids the von Neumann bottleneck as Non-Von had done. The

Computing in Neutral Territory

The neutral territory (NT) algorithm computes the forces on each atom, as determined by nearby atoms. The maximum distances involved are on the order of 10 angstroms (an angstrom is about 10^{-10} meter.)

Suppose we represent A's location by its spatial *x*-, *y*-, and *z*-coordinates—call them xA, yA, and zA. Similarly, we repre-

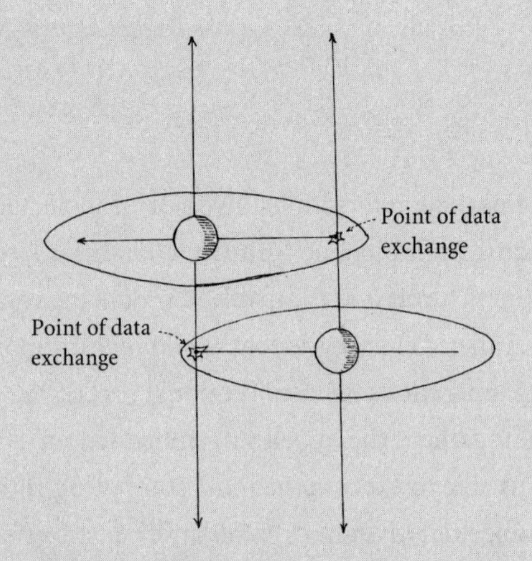

Each atom sends its information out on a 10-angstrom circle in a plane defined by its z-value (its vertical value; so the plane is horizontal and extends over x- and y-values) and 10 angstroms along the column defined by its x- and y-values. When the information about two different atoms meets, the force effects of one atom on the other are calculated.

sent B's location by xB, yB, and zB. In the NT algorithm, A sends its information across the plane defined by its height (the zA value) up to a distance determined by the notion of "nearness" it uses. A also sends its information through the column defined by its xA and yA points, up to the same maximum distance. B similarly sends information across the plane defined by zB, as well as the column defined by xB and yB. When the A and the B information meet at the intersection of the plane and column—xA, yA, zB—the effect of A on B is calculated. When the A and B information meet at the point xB, yB, zA, the effect of B on A is calculated.

result is that a millisecond of protein dynamics may be simulated in mere days, as opposed to 100,000 years. Even compared with other 512-processor devices using similar fabrication technology, Anton is 1,000 times faster. Design can make a difference. "It's not a general-purpose supercomputer, and it wouldn't be especially good at executing random scientific applications, but it's really good at what it does," says Shaw.

Other parallel machines can simulate the motion of a large (24,000-atom) protein-water complex at a rate of about 100 nanoseconds per day. By contrast, Anton will be able to perform on the order of 10 microseconds of simulation per day—roughly 100 times more. With his characteristic modesty, Shaw eschews the horse race nature of such comparisons, noting merely that specialization has its benefits.

Shaw's target application is computer-aided drug design.

A typical example might be to understand how a cancer drug affects a protein that causes cancer. One of the proteins that Shaw and his team have studied causes a form of leukemia when it's mutated. The mutation causes the protein to get stuck in an active state, thereby allowing cells to replicate uncontrollably. A relatively new cancer drug extends the lives of many patients by locking the protein in its inactive state. "We wanted to understand at an atomic level what causes the protein to switch between its active and inactive states, and to see how the protein moves and changes shape when it does. If we can understand these sorts of things at a detailed level, someone might someday use that understanding to save people's lives," says Shaw.

Just as the historical Antoni van Leeuwenhoek was able to use his powerful new microscopes to make the first observations of muscle fibers and single-celled organisms, so might Shaw's Anton machine offer new insights into the interaction of proteins, the working of existing drugs, and the design of new ones. Although drug design is a long-term goal, Shaw's immediate aim is to advance the state of scientific knowledge. He wants his lab to be a small, specialized version of the old Bell Labs. He wants to publish basic scientific discoveries in peer-reviewed journals to share the insights and to receive feedback from the community.

Along the way, his team will run superlong simulations to test fundamental working hypotheses about the accuracy of molecular dynamics simulations. These simulations are based on current models of interatomic forces. Some researchers believe that the classical Newtonian models will be able to reproduce a wide range of biologically important phenomena once it's possible to simulate molecular dynamics for longer periods of time. But the possibility remains that without the incorporation of quantum mechanical effects or other physical subtleties, very long simulations may eventually go astray. Some inaccuracies in the calculations might accumulate over the trillion steps.

On January 26, 2009, Shaw's group unveiled Anton to the scientific community. The size of a large walk-in closet, Anton resides on a raised platform in a specially cooled room. Its 512 special-purpose processors and the cooling system use as much electricity as 30 homes. A live demonstration showed a movie in which successive frames simulated 10 picoseconds (trillionths of a second) of a protein being folded in water. Every 100 picoseconds of the folding process, the protein flipped and changed shape. It was possible to trace the path of pairs of ions as they traveled through the structure. The chemists working with Shaw say they can back up a calculation at any point and step through it picosecond by picosecond. This is molecular dynamics as never before practiced. Please welcome Anton, the computational femtoscope.

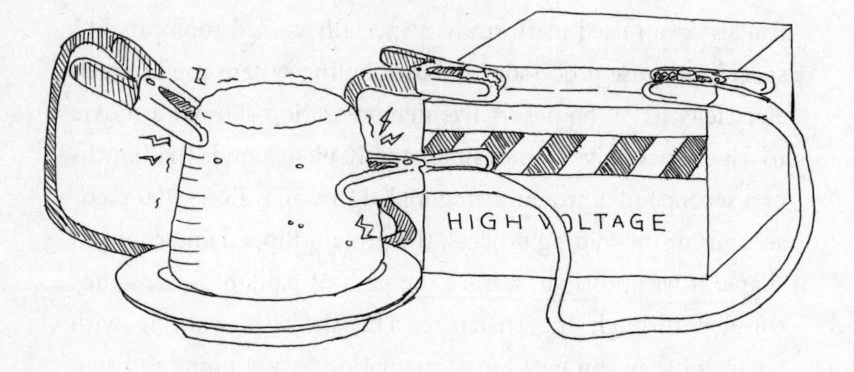

■ ■ ■ ■

I don't think like other computer scientists. I think visually; I think kinesthetically. I have motion dreams. I will see things happen and I will become an analog process.

— *Jonathan Mills*

Chapter 13

JONATHAN MILLS

■ ■ ■ ■

Doing What Comes Naturally

SCIENTISTS OFTEN MODEL PHYSICAL PHENOMENA WITH DIFFERential equations, which characterize change according to a current state. To take an example from economics, the change in the quantity of money you have in your bank account is proportional to the amount in your account—your current state. If you begin with $100, you will enjoy an increase of $5 by the end of the year, assuming a 5% annual rate of interest. If you have $1 million, you will enjoy an increase of $50,000 after the first year. This rich-get-richer relationship is reflected by any differential equation of the form $dy/dt = ky$, provided k is positive.

The world of differential equations can look a little frightening to the uninitiated but, for our purposes, you need to know that there are two kinds: ordinary differential equations and the more general partial differential equations. *Ordinary differential equations* can describe Newton's laws in physics. (In fact, Newton invented calculus, which includes differential equations, to describe his laws.) *Partial differential equations* model electromagnetism, heat and electron diffusion, quantum mechanics, and stock options pricing.

Suppose you want to solve a differential equation problem. You want to find the temperatures across an iron bar where one side is touching ice and the other is touching boiling water. One approach is to figure out the differential equation that models this situation, program it on a digital computer, and come up with a prediction. This digital approach follows the paradigm illustrated here:

question → model → program → calculate → answer

Another approach is to set up an *analogous* physical system, preferably electrical for the sake of speed and neatness. This analogous system would *measure* the electric charge everywhere and infer from those measurements the temperature of the iron bar at every point. This analog approach follows a different paradigm:

question → analogy → measure → answer

If it could be made to work, the analog approach might be far simpler and faster than the digital approach. Describing a natural phenomenon requires a differential equation; measuring the phenomenon tells us how the differential equation behaves. Suppose there are two physical phenomena: X and Y. Let's assume that the measurements of X equal the results of calculating the solution to D, a differential equation. If D models another phenomenon, Y, then why not *measure* X when trying to predict Y instead of programming D to *calculate a prediction for* Y? The analog approach is the computational equivalent of cutting out the middleman.

This idea predates the digital computer age. In 1845, one of the first scientists to study partial differential equations, Gustav Kirchhoff, used a copper plate to study how heat flows in metal. About a hundred years later, Vannevar Bush developed a "differential analyzer," using gears, curve plotters, and cams to study problems mechanically. (Bush is the visionary credited with the hyperlink concept that is so fundamental to the World Wide Web.) Bush's machine and its electronic successors could solve ordinary differential equation problems before digital computers even attempted them.

Programming electronic (mechanical) analog computers was very hands-on. A scientist would plug wires into various holes of a board (a *patch* board), carry the board to a lab, and then connect the patch board to a device, turn on the power, and read the result a few seconds later. The flow of electricity through amplifiers, resistors, and wire junctions gave the result. These devices, called *general-purpose analog computers*, could handle arithmetic and ordinary differential equations, but not partial differential equations.

The *extended analog computer* adds a solver of partial differential equations to the analog computer. Lee Rubel, the mathematician who came up with the idea, was particularly interested in partial differential equations because of his fascination with how the brain works. Neuronal research had shown that brains solve the equivalent of partial differential equations in order to model space and time. Rubel designed a conceptual model of his machine, but he thought that his machine could never be built. Fortunately, he was wrong.

Jonathan Mills and Bryce Himebaugh of Indiana University, Bloomington, found a way to build several generations of Rubel's machine. They constructed a collection of machines that implement partial differential equations in resistive sheets made out of silicon, foam, or even Jell-O. Programming the most recent prototype consists of delivering current to one of 25 points on the sheet (current *sources*), taking current from those points (current *sinks*), and then measuring voltage or feeding an output circuit. These functions take the results of the sheet to generate a visual representation of the computation or to compare the output of the sheet with a known answer. Mills and Himebaugh then readjust the current sources and sinks to obtain a better answer or to feed the output into another sheet. If this seems like a different way to think about computing, Jonathan Mills is a different kind of computer scientist.

Mills was born in Kalamazoo, Michigan, in 1952. "Do not sing that song. I can see that you're thinking about it," he says, referring to the famous Glenn Miller tune "I've Got a Gal in Kalamazoo" (with the refrain "zoo, zoo, zoo"—so there!). His paternal relatives were Cherokee, English, and French. His Uncle Lex built everything from scratch on his farm, including a tractor made out of spare car parts and I-beams that he welded together. "It was a strange-looking, H-shaped device, and he had mounted a rocking chair on it so he could ride—he didn't rock—he said it was more comfortable," Mills recalls.

Mills's grandfather had been a Baptist pastor and had wanted his son to follow him into the clergy, but Mills's father decided to join an engineering firm. He became a quality control manager for Stryker Medical, a company that built arthroscopic surgical tools and other medical devices. Arthroscopic surgery depends on a small, snakelike device called an *arthroscope* that is inserted into the patient's body and enables doctors to see the affected region using a tiny camera. When one model of the Stryker devices first went out into the field, they were immediately returned as unusable. Mills's father figured out why surgeons who had tried to use the device had failed. At a meeting of engineers and the production staff, he explained the problem. He took two empty tomato soup cans. One of them had a saline solution used for wound irrigation; the other one was empty. Both cans had a dime at the bottom. Mills Senior challenged the engineers to pick up the dime using the arthroscope. They could pick up the dime from the empty can, but not from the one with the saline solution. In designing the camera, the engineers had forgotten that light refracts when it goes through a substance other than air. They had never tested their design in solution.

Later, Mills Senior worked for Gibson Guitars. It was his job to maintain the microwave ovens used to heat the guitars. He was also an avid radio hobbyist, designing antennae to pick up aviation signals from as far away as Patagonia and Siberia.

Mills's mother worked at a school for the mentally handicapped with children as old as eighteen who had the mental capacity of two-year-olds. To help the kids, she wrote hundreds of songs that the patients could learn to sing. "Mom made them songs about life, their life—what they saw around them," Mills recalls.

By kindergarten, Mills was already reading adult books. In 1950s Michigan, there were no schools for talented children. Mills's mother requested tests to see whether he could skip a grade, and he ended up in third grade, smaller and younger than his classmates by a year. His bigger problem, though, was with his teachers. In third grade, the class studied a unit on rockets. Mills had a book called *All about Rockets and Jets.* It explained the story of Cape Canaveral and the invention of jet engines and rockets. The teacher told the class that countdowns started at 10. Mills's hand shot up to correct the teacher. He said that countdowns started at 100 or more. "The lightning began to play around the brow of this teacher. So she said, 'You come up to the front of the class and you tell us all how it goes.' I talked until I cried," Mills recalls.

As he grew up, Mills built devices of various sorts. He also collected insects and learned how to raise butterflies. His mother would drive him along country roads and he would pick up butterflies that had been hit by cars. At Western Michigan University, he majored in Latin, which enabled him to be in small seminars with graduate students. By contrast, the chemistry department had classes of three hundred.

When he ran out of money, Mills decided to take a scholarship from the US Army's Reserve Officers' Training Corps (ROTC). The United States was fighting the war in Vietnam, and Mills was going to be sent in country through Cam Ranh Bay. But then President Nixon started pulling out troops, so Mills was sent instead to Dugway Proving Ground in Utah, to study captured Soviet chemical equipment. His boss, Dr. Robert Stearman, discouraged Mills from making a career in the army. Instead, he gave him a key to a building with a Wang word processor and a DARPAnet link to White Sands, New Mexico, where there was an IBM 360. (DARPAnet was the Defense Advanced Research Projects Agency's precursor to the Internet.) Stearman said he would teach Mills to program. "Well, he was offering candy to a baby," says Mills.

Mills wrote a program to figure out how to schedule the tasks required to decontaminate vehicles. He rearranged the workflow so that each vehicle required 15 minutes to decontaminate instead of an hour. The downside was that the soldiers were working continually. Mills's superiors liked the efficiency, but the soldiers had enjoyed a more leisurely pace when decontamination took an hour. Mills felt their resentment, but he had found a calling: "Computers were the closest thing to magic. When you program a computer, you touch something that is invisible; it's powerful, it's controllable, and it's flexible."

Mills next went to the Argonne National Laboratory. Founded in 1946, Argonne was the first national laboratory and is now located on a campus twenty-five miles from downtown Chicago. There, Mills built a machine for the logic language Prolog. For several months, he had the world's fastest Prolog compiler. On the strength of that work, Motorola funded his doctoral studies at Arizona State University. His thesis was on logic programming.

In *logic programming*, a programmer states constraints on a solution instead of how to arrive at the solution. For example, given parent-child data, consider finding all the ancestors of an individual X. In most programming languages, you would write a program that would find the parents of X, then find the parents of the parents of X, and so on, until there were no more parents to find. In Prolog/logic programming, you define an ancestor to be either a parent or the parent of an ancestor. The interpreter (what Mills built) would find the set of ancestors.

Mills published his thesis and then decided to become a professor—a lifelong dream. He joined the faculty at Indiana University, Bloomington, and began his work on analog computers. It makes sense in retrospect—he had a strong background in physics and chemistry, and his Prolog work had led him away from thinking about computation as a step-by-step recipe and more about thinking about computation as a way to specify a solution.

Mills felt that technology was on his side. He imagined linking 64,000 processors together. Danny Hillis, the founder of Thinking Machines, a computer manufacturer that produced a neuronally inspired series of machines, had built a machine to try to simulate the workings of the human brain. Interconnecting the processors required tremendous quantities of wire. To digital designers, this was pure overhead. But not to Mills: "I asked, how do you make wires compute?"

Crazy questions sometimes lead to promising paths. Adding two 64-bit numbers on a digital computer requires hundreds of transistors, resistors, and little pieces of wire. But now assume that you drive current X down one wire and current Y down another wire, and the two wires meet a third wire at a junction. How much current goes down the third wire? You probably guessed it: X + Y. Even if you don't feel so comfortable with electricity, you've seen this if you've ever seen two river tributaries come together. Where they join, a new river flows that has the water from both rivers. So, three wires can add. That replaces lots of circuitry.

In 1993, Mills became intrigued by some theoretical papers by a man named Lee A. Rubel, a mathematician at the University of Illinois at Urbana-Champaign. His first contact with Rubel was via e-mail. The next was a "package of his papers that came in a brown paper envelope. They reeked of cigar smoke," Mills recalls. Among the papers was Lee Rubel's vision of the extended analog computer. In its ideal form, it could compute many more functions than the earlier analog computers—notably, partial differential equations.

How exactly do analog and digital machines differ? Let's start with the basics. A digital computer decomposes all of its data into tiny pieces called *bits.* A bit is either a 0 or a 1, so a digital computer has to approximate values such as pi (π). An analog computer stores values as voltage levels or currents, which means it can hold any real-number value. Mills says it's as if the digital computer has to build everything it uses out of Lego blocks, while the analog computer builds things out of Play-Doh. The blocks come with detailed instructions about how to put them together to make something like a toy airplane. The clay doesn't have any instructions; if you want to make a toy airplane, you remember what it looks like and then mold and sculpt the clay to match your memory or a picture of the airplane.

Each machine has its best applications. Digital computations excel at processing text, storing addresses and personnel information, and accounting—the vast bulk of computing known as *data processing.* Analog computing could be best for finding visual patterns, modeling physical processes, and controlling machines. The two kinds of computers treat the underlying physics of computer circuitry completely differently. The digital computer takes devices like transistors and resistors and forces voltages into 1s and 0s, after which the underlying physics is ignored by the programmers and users at higher levels. By contrast, an analog machine uses the full range of values offered by the underlying devices, potentially permitting much more computation from fewer devices.

A digital-circuit designer might reply that cramming more devices onto chips is now easy, but figuring out what to do with them is hard. This is true up to a point, but the chief obstacle to speed in a digital computer is the need to fetch data from memory and store it back in memory after doing a fairly simple operation, such as adding two numbers (the von Neumann bottleneck discussed in the chapter on David Shaw).

By contrast, an analog computer could allow one operation to flow into the other. Better yet, the underlying physical device can be made to *embody* the same mathematical model as the system under study, thus completely avoiding the need to solve multiple differential equations, at a cost of billions of additions and multiplications.

Principles of Analog Programming

It is a tradition among computer scientists when becoming familiar with a language or new computer to write a program that prints the sentence, "Hello, world." When Mills asks his students to create an *H*, at first they think digitally like a dot matrix—points in the form of an *H*. "An analog way of thinking is to ask whether you can put something in there to create hills and valleys. If you would look at them as a topographic map, you could create an *H*." That's the "Hello, world!" program for an analog computer.

Mills produces the hills and valleys by creating current sources at certain points and current sinks at other points. In his most recent configuration, he has 25 possible points of control. He has found a pattern of sources and sinks that would generate every letter in the alphabet. Does each letter look the same every time? Well, yes and no. He can produce an *H* that will look like an *H* every time, but it won't be exactly the same *H*.

This is admittedly a lot of work to make a simple letter. But the general approach applies to far more difficult problems. H. Fredrik Nijhout at Duke University had observed that partial differential equations with set boundary and interior conditions could generate every pattern in any butterfly's wing. Kjell Sandved had found that the wing patterns made every letter and numeral in the English alphabet. Mills combined those two observations to create a universal pattern generator for butterfly wings, as discussed in the box "Butterfly Wings, Baking Soda, and Paper Towels."

Creating the letter *H* and modeling wing patterns may strike you as trivial pursuits, but the extended analog computer can potentially serve as a supercomputer to solve Grand Challenge problems too. (*Grand Challenge problems* are those that computer scientists find hard to solve but are important to science and society.)

Mills's idea is to use the extended analog computer to propose solutions to a problem but then have a digital co-processor check those solutions. If the answer is wrong, the digital co-processor sends a feedback signal to the analog processor, causing it to readjust the values of its current sources and sinks and hopefully improve its guess. Mills first applied this strategy (proposing in analog and then correcting in digital) to the problem of constructing an artificial retina. His idea was to translate an image into electrical inputs on the conductive foam and then have that foam produce a set of voltages that could be compared to the voltage profile of known objects. The computer starts by assuming that the first object is an example of a category of similar objects. When the extended analog computer—the artificial retina—"sees" an object that is not in a known category, the functions of the logic gates are modified (via digital feedback) to separate the new object into a category of its own (see the box "Computing on Foam").

Butterfly Wings, Baking Soda, and Paper Towels

In 1952, Alan Turing (the father of modern digital computation) provided the first thorough description of biological pattern formation (***morphogenesis***) using reaction-diffusion equations. As the name implies, ***reaction-diffusion equations*** look at how elements interact with other entities and spread across their environment.

Using an algorithm to solve partial differential equations, a digital computer models morphogenesis. These algorithms use one or more matrices with certain boundary values. The system of equations is evaluated by manipulation of the matrices set up by the boundary conditions. Every part of the equations corresponds to a part of the algorithm. The operation of a digital computer program makes it easy to translate the matrix to elements in memory, and to translate the equations to simple arithmetic operations such as add, subtract, multiply, and divide. Repeated sequences of these operations make up the algorithm and are controlled by nested loops (loops within loops) that step through the matrix elements.

By contrast, instead of using an algorithm to solve partial differential equations and to simulate the behavior of systems, the extended analog computer uses the properties of nature itself. There is a simple experiment you can do in your kitchen to see firsthand how the extended analog computer would work for this problem. In the following recipe—by using baking soda, vinegar, food coloring, a paper towel, measuring cups, and a flat pan—you can demonstrate how butterfly wing patterns are formed.

First cut a sheet of paper towel to fit inside the flat pan. Then fill one measuring cup with a baking soda solution in water dyed with red food coloring, and another measuring cup with vinegar. Pour the baking soda solution slowly onto the paper towel. The food coloring mixed in with it will spread outward in a circle.

Because natural law governs all physical processes, the paper towel and the butterfly wing work in a similar way, but the timing may differ. In the "kitchen science" experiment, it takes seconds for the dye to diffuse through paper. But butterfly wing diffusion takes days as chemicals flow through the cells in a developing chrysalis.

In the extended analog computer, electrons diffuse through a lattice of silicon atoms in about a nanosecond. A digital computer can simulate all of these systems, but it must do so indirectly, step by step with a program. The analog system operates using the principles of nature directly. In the extended analog computer, the current sources and sinks can be placed at any of the 25 points on the grid. To model butterfly pattern generation, a current source is placed in the center with sinks at the four edges of the grid.

Note that the kitchen experiment also models the reaction-diffusion equation, by making a bubble under the paper towel. It begins when the vinegar is gently poured into the center of the pink circle. Look carefully, and you will see that surface tension holds the paper towel to the flat plate, but one or more tiny "balloons," or bubbles, develop under the towel. As the vinegar and baking soda react to produce carbon dioxide, a cir-

(continued)

(continued)

cular balloon spreads out from the center. But because the gas slowly escapes from the tiny balloon, after a few minutes you may be lucky enough to see the balloon form a circle or hollow diamond of little bubbles, depending on the kind of paper towel you used. This phenomenon is analogous to morphogenesis as described by Turing, and to butterfly wing pattern formation as described by Nijhout.

Using feedback is so central to our everyday experience that we are barely aware of it. Feedback through our eyes enables us to drive a new car down a new road without incident. Without visual feedback, we would not even consider the trip, even if we had driven down that same road in the same car for years. Dogs that catch a Frisbee use feedback to catch the flying disk even in irregular winds. In the future, using the power and speed of analog computing to propose solutions and then checking the solutions with the precision of digital computers (or even a person's senses) may appear as natural to our grandchildren as iPhones are to our children.

Mills recognizes that he thinks differently from most of his colleagues who have imbibed the digital culture. He says typical computer scientists are "very big on control. They want to specify everything in advance and control the process so that the machine works correctly." Mills studies problems by building models (for example, ball-and-stick models of molecules) and imagining how these will be implemented on circuits. "I think visually; I think kinesthetically. I have motion dreams." In his spare time, Mills integrates New Age music with phantasmagorical artwork.

Thinking differently is risky. When Mills came up for tenure, a colleague told him he was outstanding in his field. Then he added that Mills was the only one who was outstanding in his field. Mills did receive tenure, but the question remains whether his approach will catch on. Mills is optimistic about a new generation of computer scientists who might appreciate the simplicity of the analog approach. Thinking about a problem from the math and the program perspectives, he believes, is fundamentally a lot harder than his approach.

Use of a configuration rather than a program in the extended analog computer raises another issue first posed by digital computing pioneer John Backus in 1977.* A digital computer must be explicitly programmed to solve a problem. A program may cause unwanted results. Backus called this "programming in the von Neumann style" and asked if computing could ever be liberated from it. Mills thinks the answer is yes, but not with a digital computer.

Analog computers use the similarity of the physics of materials to compute. The "computing elements" in the extended analog computer's primary component (the conductive sheet) are themselves atoms or molecules. Its digital components—binary logic gates, memory, and wires—do only what they're good at. In

* For a profile of Backus, see our first book, *Out of Their Minds: The Lives and Discoveries of 15 Great Computer Scientists.*

Mills's work with New York University scientists on protein folding, digital components handle short-range interactions while analog components handle long-range ones.

In a way, this division of labor should not be surprising. Another renegade scientist, Richard Feynman, also felt that digital computation might not be the best method for modeling nature, at least not all of it. Mills says, "Remember Feynman's famous question: Why does it take a digital computer so much computation to model what goes on in a little piece of nature?"

Computing on Foam

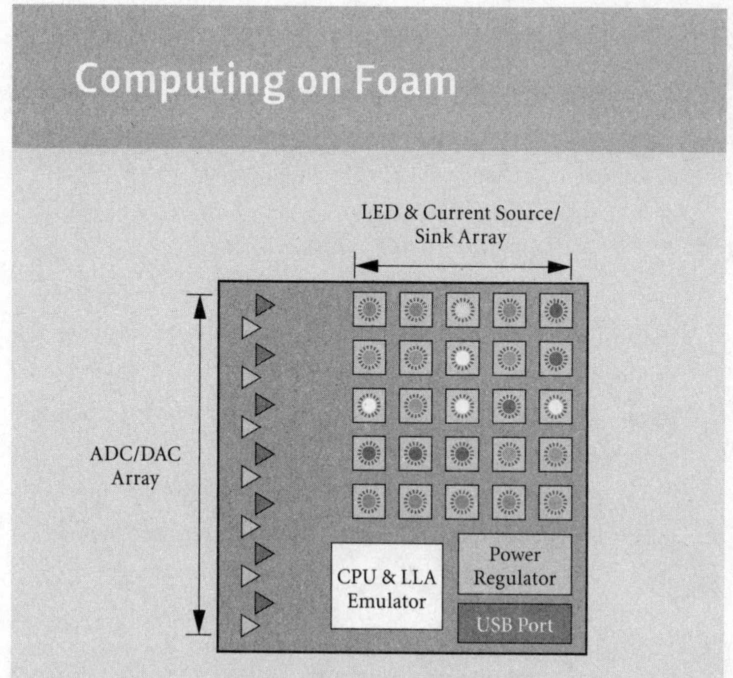

The current version of the extended analog computer: top view.

You can find the supplies to build the hardware of the extended analog computer in any traditional hobby shop. It is constructed on a single circuit board, with the top side used for the integrated circuits, and the bottom side for the conductive foam sheet. Both the top and the bottom are covered with clear Plexiglas to protect the components when they are being moved. The translucent Plexiglas square on top diffuses the light emitted from the five-by-five blue-light diode

(continued)

(continued)

array, which gives a rough indication of how the output, a voltage gradient, is changing. Each of four holes, one in the middle of each edge of the translucent Plexiglas, provides access to a small pin, visible as a dark spot (see the figure above). The pins allow one Plexiglas computing sheet to be connected at the north, south, east, and west edges, so a stack of machines can model three-dimensional systems. That's how Mills plans to configure extended analog computers for problems such as protein folding.

On the bottom side of the circuit board is a sheet of black plastic conductive foam, the same material that is used to pro-

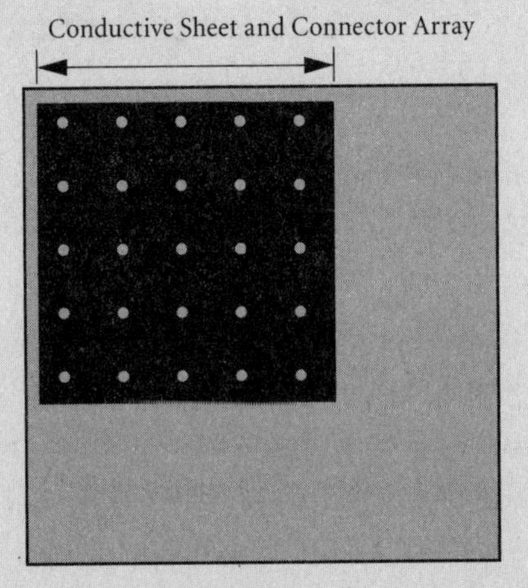

The current version of the extended analog computer: bottom view.

tect integrated circuits during shipping. (Mills jokes that when a colleague purchases a computer, he can use the packing to make one of his own.) The foam sheet is attached to the board by being pressed onto a five-by-five array of metal pins. Each pin connects to a special-purpose circuit on the top side of the board, which also includes a light-emitting diode. Each circuit may be individually configured to perform one of the following three functions: (1) supply a positive or negative current to the foam, (2) read a voltage directly from the foam, or (3) read a voltage and scale it for use with a Łukasiewicz logic function.*

The design looks simple, but the computation makes use of the physics of foam. The simplest view of the foam's behavior is that it uses Kirchhoff's current law to sum the flow of electricity from whichever of the twenty-five pins are supplying positive and negative currents. ***Kirchhoff's current law*** states simply that the sum of the current coming into a junction must also leave that

(continued)

■ ■ ■ ■

* *Łukasiewicz gates,* named after the Polish mathematician Jan Łukasiewicz, are different from digital ones. Whereas a Boolean gate (the typical digital gate) takes 0s and 1s as inputs and gives 0s and 1s as outputs, Łukasiewicz gates take continuous values between 0 and 1 as inputs and produces continuous values as outputs. For example, a Boolean inverter takes a 1 as input and produces a 0 as output, or takes a 0 as input and produces a 1 as output. A Łukasiewicz inverter takes a value x as input (between 0 and 1) and produces 1 – x as output. The benefit of a continuous circuit is that any of an infinite set of values can result.

(continued)

junction. Electricity flowing through the sheet encounters resistance, so there are measurable voltage drops. The capacitance in the sheet, although small, introduces a delay that is useful in computations involving equations for oscillating systems. The next level of modeling makes use of the fact that electrons diffuse through the continuous sheet. Because this diffusion is modeled by partial differential equations, a programmer can "solve" such equations by measuring voltages.

Other materials have been wired to the input pins too. Silicon integrated circuits act as sensors that generate voltage gradients. That phenomenon was useful for the electronic retina. Jell-O can be molded around three-dimensional arrays of contacts to model three-dimensional systems (weather prediction, protein folding, and others). The sheet enables the machine's user to directly access the properties of nature as its fundamental "instruction set." Mixing and matching different materials increases the computing power of the extended analog computer.

The extended analog computer begins operation in a digital host computer (say, a conventional personal computer) that sends data through a USB port to the circuit board of the extended analog computer. There, the digital signals are converted to analog and cause currents to flow through some of the pins in the five-by-five array under the foam. Other pins in that array measure the voltages and transmit those values back through continuous Łukasiewicz logic gates to the host digital computer, where they can be manipulated via standard programs.

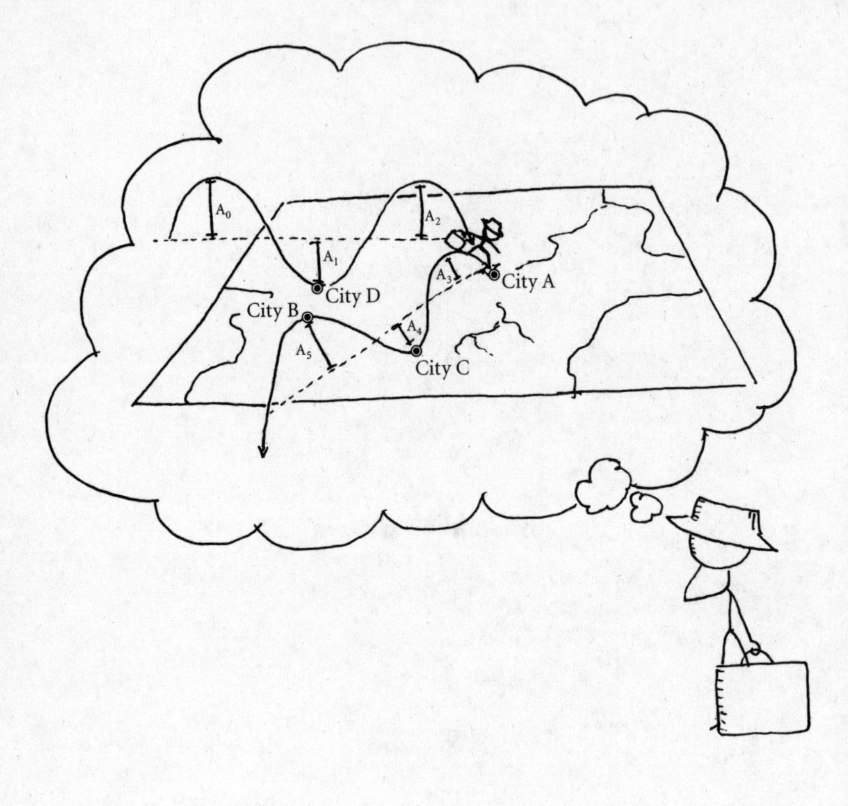

■ ■ ■ ■

For a century, physicists did a very good job of making everyone believe that quantum mechanics was hard. And then I found that quantum mechanics, once you take the physics out, is just a version of probability theory with minus signs. That was a revelation. I don't really know physics, but I can multiply matrices and vectors.

— *Scott Aaronson*

Chapter 14

SCOTT AARONSON

■ ■ ■ ■

Finding a New Law of Physics

AS OF THIS WRITING, NOBODY HAS BUILT A QUANTUM COMPUTER having more than a handful of quantum bits. Yet celebrated theorists have designed remarkable algorithms for quantum computers. Others have explored the theoretical limits of these largely theoretical machines. Secretive government agencies pour money into their construction. Why so much attention to something that is still so speculative?

The motivations vary. Spies are interested because remarkable algorithms suggest that quantum computers may be able to break codes exponentially faster than conventional computers can. Some physicists believe that quantum computers (if built) would demonstrate the existence of millions of parallel worlds. Mathematicians think that quantum computing offers a window into understanding the hierarchy of problems known as *computational complexity.* Within the last group is Scott Aaronson, a mathematician whose work in complexity theory led to an interest in quantum computers. On the wall of his office at MIT, Aaronson has a poster that a friend gave him for his birthday called "The Complexity Zoo." He is working hard at trying to tame its inhabitants.

Born in 1981 in Philadelphia, Aaronson grew up in Washington Crossing, the town where George Washington led his army across the Delaware River to fight the Hessians in the Revolutionary War. Aaronson's mother was an English teacher. His father worked as a science writer at Bell Labs when Arno Penzias and Robert Wilson won the Nobel Prize for their discovery of cosmic microwave background radiation. Aaronson senior helped describe the discovery to the public.

At an early age, while other kids played basketball, Aaronson thought about extreme scientific challenges, such as the speed of light. "The idea that there was this fundamental limit was the thing that interested me; also designing spacecraft to go at the speed of light. I don't think I understood that if they had mass, they couldn't do that," he recalls. Aaronson also became enthralled by the virtual worlds he could create on a computer. He would wander around by himself thinking about making his own video games. "It just seemed like entire universes that you could create. I didn't understand how it happened. Maybe thousands of people in white lab coats somehow assemble a video game like you would assemble a 747 airplane," he remembers thinking.

One day when he was eleven, a friend showed Aaronson a space game on his Apple II. The friend had the code for the game. He explained how changing a line would get the game to do something different. "It was like finding out where babies come from. I was angry and upset that nobody had told me about it earlier," says

Aaronson. Then he took it one step further. Instead of programming, he wondered if he could create his own programming language that would let him do things beyond Apple BASIC. "I still didn't have the idea that programming languages are all the same slop—the Church-Turing thesis," admits Aaronson.

The *Church-Turing thesis* asserts that no digital computer is more powerful than what can be expressed by a simple programming language with instructions to do arithmetic on whole numbers, to loop over a sequence of instructions, and to stop when the computer memory has a certain value. Any language can make complex computations easier to express, but it doesn't fundamentally change what one can do with a computer.

Aaronson spent the next few years doing experimental computer science in his free time. He created cellular automata and fractals (geometric patterns that repeat themselves in ever-smaller copies) and watched them move around. Aaronson's futzing at age twelve covered some of the same ground as Stephen Wolfram's *A New Kind of Science*, published in 2002. Unlike Wolfram, however, Aaronson never did any experiments. "I don't cook. I'm not good at working with my hands," he says. Instead, he focused on the limits of computer programs—what they could or couldn't do.

Finding school's limitations was far easier. "I felt that if I was going to be in what amounted to a prison, then I should be sentenced by a judge," he recalls. Aaronson was bored and restless in school. When his father took a job with AT&T Asia/Pacific and the family moved to Hong Kong, Aaronson became even more frustrated with school. He was told that high school would be better, so he skipped a grade and moved on to high school. "I once read that if you have an animal in a cage and you open the cage, it stays there for a while until it gradually dawns on it that it can get out. It gradually dawned on me that I should just get out of high school as quickly as possible." Aaronson returned to the States and ended up skipping another grade. But he didn't want to spend another year in high school. There wasn't enough sophisticated math.

All told, Aaronson skipped three grades. He then decided to switch to a high school equivalency program at Clarkson University in upstate New York, which was known for transitioning high school kids to college. Aaronson is still haunted by what he might have missed in high school. On his website, he describes in a posting called "The Beehive" returning to his high school, where he is a solitary bee and imagines a counterfactual scenario in which he would have been engaged in the buzz of gossip, activity, and adolescent ritual.

The summer before entering Clarkson, Aaronson attended a math camp at the University of Washington in Seattle, where he was introduced to concepts that became relevant to his life's work. "It was a revelation for me," he says. He heard Richard Karp give a talk about complexity theory. A Turing Award winner, Karp had helped develop the field of computational complexity, particularly by showing the pervasiveness of NP-complete problems (see the box "The NP Completes").

While obtaining his high school equivalency degree, Aaronson wrote his first scientific paper, under the mentorship of Chris Lynch, a computer scientist at Clarkson who specialized in *automated theorem proving*, the field that uses a computer to prove mathematical theorems. Aaronson graduated from Clarkson and went on to Cornell to work with Bart Selman, who was the advisor for a robotic soccer player project. "That convinced

The NP-Completes: A Family with Problems

P stands for ***polynomial***. A polynomial equation in *n* has variables and constants, as in $5n^3 + 2n - 15 = 0$. ***NP*** stands for ***nondeterministic polynomial***. A computational task (a "problem," in computer science parlance) is ***in P*** if there is an algorithm that can solve it that takes time proportional to a polynomial in *n*, where *n* is the "size" of the problem.

For example, finding out whether the number 873,895 is in an unordered list of 10,000 numbers is a problem of size 10,000 ($n = 10{,}000$) and takes time proportional to *n*. Other problems, such as aligning two DNA sequences of size $n = 10{,}000$ require time proportional to n^2. A problem that takes time proportional to n^2 or time proportional to n^{20} is in P, for the simple reason that *n* (the size of the problem) is in the base of the time expression rather than its exponent.

If a problem requires trying all possibilities (what Russian mathematicians colorfully call ***perebor***, or "brute force," problems), then the time required may be exponential in *n* (that is, *n* may be in the exponent of a particular expression). For example, in the oft-cited traveling-salesman problem, the time required to try all possible routes through a group of cities to find the cheapest route that goes through them all is roughly proportional to n^n. Because there is an *n* in the exponent, this is exponential time.

NP problems have solutions that can be ***verified*** in polynomial time. If I give you a route through cities and claim that

(continued)

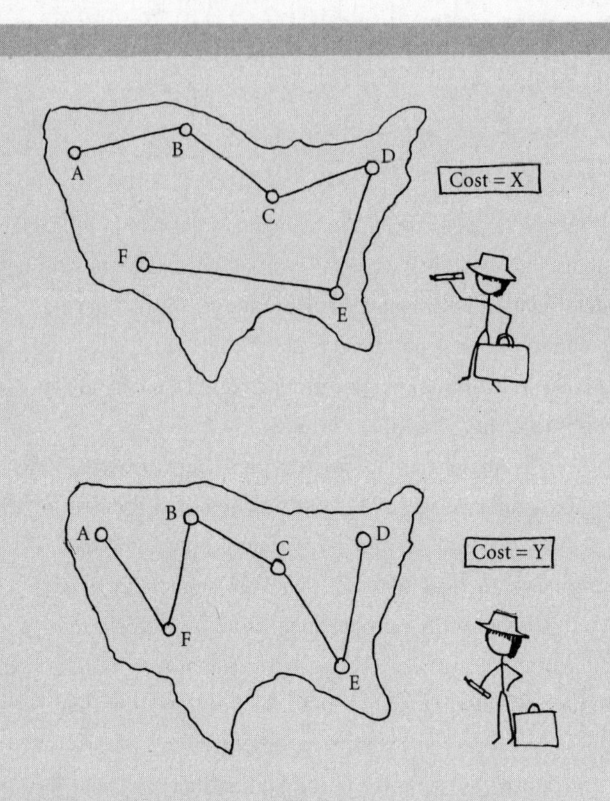

The traveling-salesman problem. Two possible routes for the salesman are illustrated. Given any route, one can easily compute its cost. The problem is to find a route where cost is below a certain budget.

(continued)

the route has a certain cost, it takes time proportional to ***n*** to verify that claim. Operationally, a problem is ***NP-complete*** (as opposed to in P) if the only known methods for ***finding*** solutions to the same problem take exponential time.

Exponential time quickly becomes impractical. For example, going through every route for 100 cities would require more than a billion processors working for a billion seconds. For that reason, the pragmatic strategy for such problems is to use ***heuristics***, fast techniques that may not give the best possible solution, but often give very good ones. Between NP-complete and P problems lies a nether region of problems for which no polynomial time method has been found. It is unknown whether these problems are in P or are NP-complete.

The most famous problem in the nether region is factoring. Its fame comes from its importance in cryptography. You may remember from your school days finding the prime factors of a number. For example, the prime factors of 15 are 5 and 3. The prime factors of 221 are 17 and 13. The naïve method for finding the prime factors of a number *n* is to try to divide *n* by all numbers less than *sqrt*(*n*) to see if they are factors. Even if each test took only a billionth of a second, finding the prime factors of a 100-digit number could take billions of years. The time is exponential, not in *n* but in the number of digits of *n*, denoted *d*(*n*). Faster methods using "sieves" can reduce this complexity substantially, but they leave *d*(*n*) stubbornly in the exponent.

me I should not be an engineer," says Aaronson. Making his code work with other people's code was always going to be something other people were better at, he decided.

During the summer breaks from Cornell, Aaronson worked at Bell Labs when it was still, as then Vice President Gore put it, "the crown jewel" of US research. Aaronson spent a summer working for Lov Grover, one of the principal algorithm inventors of quantum computing. Even though no sizable quantum computers had been built, the basic theoretical model was already known. Aaronson was tempted by the chance to work on a new computational model: "The floodgates had been opened. People were trying to see what the scope of quantum algorithms was. How could I not be interested?"

He was in good company. Physicist Richard Feynman was one of the earliest people to think about quantum computing. Physicist David Deutsch, who is still active in the field, was another. At the University of California at Berkeley, Aaronson's doctoral advisor, Umesh Vazirani, led the first systematic studies into the inherent difficulty of problems in a quantum-computing setting.

Quantum Mechanics Essentials

Aaronson has adopted a mathematician's approach to teaching quantum mechanics. Physics courses tend to start with the experimental observations of the early 1900s and then progress through a series of approximate theories developed to explain these observations. Finally, the physics course reaches its conceptual core. Aaronson's approach is different. He tells his students, "Here are these new entities called amplitudes; let's see where they lead." Says Aaronson, "Once you take the physics out, it's a version of probability theory with minus signs."

Let's try to understand this approach step by step. Consider all possible mutually exclusive outcomes of all rolls of a pair of dice. Each possible outcome has a probability between 0 and 1. The sum of all those probabilities must add up to 1. Quantum mechanics requires us to think of an alternative to probability called an *amplitude.* Suppose an amplitude can take a value between –1 and +1 and the sum of the *squares* of the amplitudes of all possible mutually exclusive outcomes equals 1. The square of the amplitude of an outcome therefore corresponds to its probability.

What does the notion of amplitude bring to the party? Amplitudes can be added to one another, and since amplitudes can be negative (technically, they can even have imaginary components, but that's not important for this discussion), two amplitudes may cancel one another out—what physicists call *destructive interference.* Noise-canceling headphones work by destructive interference also, though of sound waves rather than quantum waves. Destructive interference leads to some strange effects.

The usual example given is the double-slit experiment. You shoot a single photon through a first screen with two slits in it. What you see on a second screen is that light never reaches certain points. But if you cover one of the slits, then the photon can hit those points. That result seems to fly in the face of the conventional understanding of probability, in which one would assume that opening a second slit can only add more possible destinations on the second screen.

The amplitude notion gives a way to predict this effect mathematically. Opening both slits means that the amplitudes from the two single-slit experiments are added. In certain places the waves of amplitudes sum to 0 and we get a dark spot. "When people talk about all the mysteries of quantum mechanics, it's really that one thing—the interference of positive and negative amplitudes—that shows up in a thousand different guises," says Aaronson.

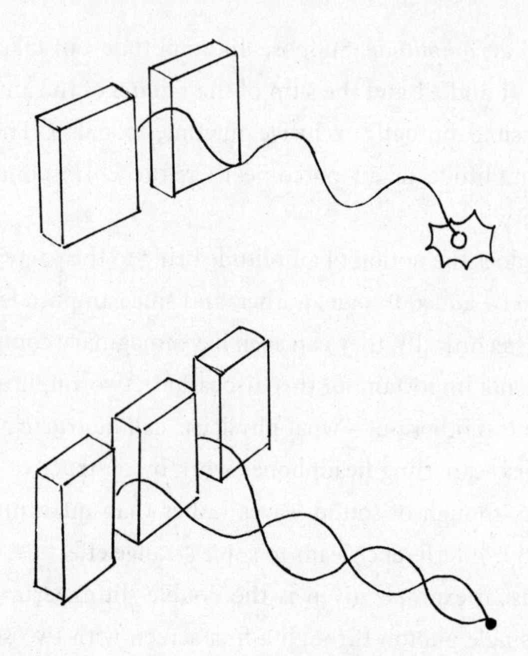

The double-slit experiment. If there is only one slit (top), then there will be at least some light everywhere, though with varying intensities. Intensity is the square of the amplitude. With two slits (bottom), the photon acts like a wave and there will be interference among the amplitudes, leading to dark spots.

Quantum Computing

Understanding amplitudes as values that can range from –1 to +1 helps us understand the notion of a *quantum bit*, or *qubit*, in a quantum computer. In a classical computer, a bit can have a value of either 0 or 1. In a quantum computer, a qubit has an amplitude at 0, denoted A0, and an amplitude at 1, denoted A1. If there are two qubits, we need to assign an amplitude to each of the four possible settings of two bits: A00, A01, A10, and A11.

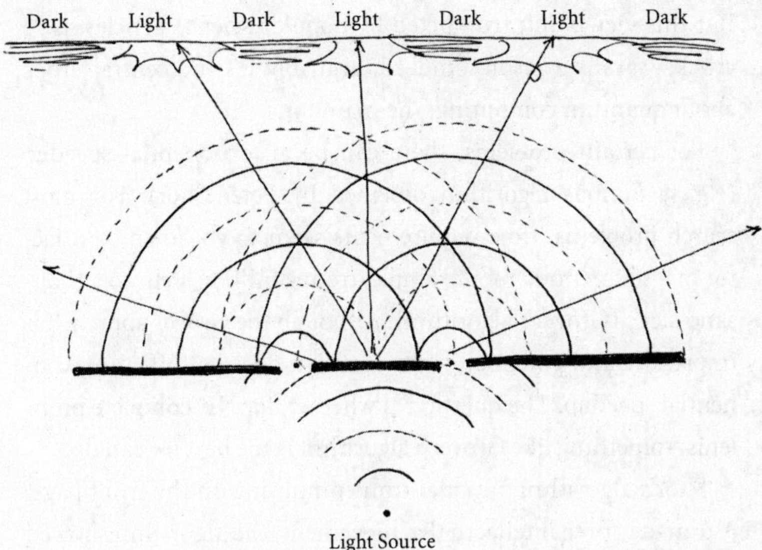

Solid lines correspond to positive amplitudes; dotted lines, to negative amplitudes. When positive and negative meet, there is darkness.

(By contrast, on a digital computer each bit has all its amplitude at either 0 or 1, so the combined state of two bits has just one nonzero amplitude.) Likewise if there are three qubits, then we need to assign the following amplitudes: A000, A001, A010, A011, A100, A101, A110, and A111.

Each time a qubit is added to the mix, the number of amplitudes that characterizes the combined state doubles. That means that the combined state of k qubits requires 2^k amplitudes. Suppose that for a problem having 2^k possible values and one "best" one, we could assign each value to an amplitude using k qubits. If we could fish out the best value from the exponential possibilities, we would have found the holy grail of exponential speedup.

But this viewpoint, trumpeted by popular science articles, is "a crock," says Aaronson. "Indeed, arguably it's the central crock about quantum computing," he maintains.

For certain problems, there can be an exponential speedup (e.g., a famous algorithm designed by Peter Shor). For most search problems, time requirements seem to go down by a factor of a square root—for example, from a million steps to a thousand steps (Grover's algorithm). Although the Grover approach is impressive and potentially very useful, it does not offer an exponential speedup. The question is whether, for NP-complete problems, something like Grover's algorithm is the best we can do.

Shor's algorithm put quantum computing on the front page. A tour-de-force intellectual achievement, the algorithm showed that factoring could be done fast on a quantum computer. Recall that *factoring* finds the prime factors of a number. For example, the prime factors of 21 are 7 and 3. Factoring's difficulty underlies the cryptography that makes your online purchases secure.

Shor's algorithm for factoring used time roughly proportional to the square of the number of digits of the number to be factored. So, for a 1,000-digit number, this would take "only" something like a million steps. This is much better than the best-known classical algorithm, which would take time proportional to 10 million billion steps. We say "known" because if any cryptographic agency found a better way to factor, that agency could break many codes but would probably not break its silence about having cracked those codes.

Shor was able to construct his algorithm by reducing the factoring problem to finding the "order" of a repeating sequence—a problem particularly well suited to quantum computing (see the box "Shor's Algorithm").

Shor's algorithm raised hopes for an exponential breakthrough, but it worked for only a very special problem. Grover's algorithm gives less of a speedup but works for general black-box kinds of problems. Suppose you want to find out whether a particular value is present in a collection of *n* values. With a classical computer, this task would require looking at roughly half (*n*/2) elements in the collection. On a quantum computer, this can be done in *sqrt*(*n*) steps.

The approach again exploits the peculiar nature of amplitudes. Grover described an operation to manipulate the amplitudes so that if the value you are looking for is present and corresponds to a bit combination, then certain operations on the amplitudes will magnify the amplitude corresponding to that bit combination. It's as if the operation rotates the combined state of the qubits toward the desired bit combination. After approximately *sqrt*(*n*) rotations, the combination is found.

A proof by Charles Bennett, Ethan Bernstein, Gilles Brassard, and Umesh Vazirani showed that each query can help only so much. For such unstructured problems, *sqrt*(*n*) is the best that can be done. If a quantum algorithm for NP-complete problems existed, it would have to exploit their structure—just as Shor's algorithm exploits the structure of the factoring problem. Unfortunately, general NP-complete problems don't seem to have enough structure to exploit.

Aaronson explains this glitch with the example of a map. Suppose you're coloring a map with three colors and you want no neighboring countries to have the same color. You don't know if

Shor's Algorithm

Suppose the problem is to factor a large composite number N. A composite is the product of two primes. Remember that primes can be factored only into themselves and 1. Quantum computing does what essentially is a really fast subroutine. It finds the "order" of A with respect to N, where A is a randomly chosen number that is relatively prime to N.

Two numbers are ***relatively prime*** if their greatest common divisor is 1. For example, 4 and 15 are relatively prime. (If A is not relatively prime to N, then they have a common factor. That common factor is a factor of N, which would give a solution to our problem, so let's assume we're not that lucky.) The order is the smallest r, such that $A^r \bmod N = 1$. (The "mod" operator is just the remainder after division. For example, 13 mod 5 is 3 because if you divide 13 by 5, you get a quotient of 2 with a remainder of 3.) It turns out that if r is an even number, then very often the greatest common divisor of N and $(A^{r/2} + 1)$ will be a factor of N.

Let's see how this would work for the very simple composite 15. We take A to be 4. As it happens, $4 \times 4 \bmod 15 = 1$, so the order $r = 2$, an even number. Therefore, $(A^{r/2} + 1) = 5$ has a good chance to be a factor of 15, and it is! Here's another example: Let A be 8. This implies that $r = 4$, because $8^4 = 4{,}096 = (273 \times 15) + 1$, so $8^4 = 1 \bmod 15$. Because $r = 4$ and 4 is even, we calculate $(A^{r/2} + 1) = 64 + 1 = 65$. The greatest common divisor of 15 and 65 is 5, again a factor of 15. This method can be used to find the factors of numbers that are hundreds of digits long.

If you are interested in more details, start with Aaronson's own description on the Web (www.scottaaronson.com/blog/?p). A physico-philosophical account is given in David Deutsch's book *The Fabric of Reality*.

there is only one way to color the map, a million ways to color it, or no way at all. By contrast, with factoring you might not know what the factors are, but you know that every positive integer has a unique factorization. Factoring enjoys many such favorable properties because of the structure of the positive integers. "This [distinction in underlying structure] is important," says Aaronson. "Many of us don't believe that there is a fast quantum algorithm to solve more 'generic' search problems," he concludes.

Quantum computers could give us an exponential speedup for factoring and a square-root speedup for unstructured problems. NP-complete problems have more structure than black-box problems, but not as much as factoring. Can we do better than a square-root speedup? Nobody knows, but there is a lot at stake. "The ability to solve NP-complete problems would give us godlike powers," Aaronson believes. "When we talk about NP-complete problems, we're not just talking about scheduling airline flights (or for that matter, breaking the RSA cryptosystem). We're talking about automating insight or intelligence."

"Insight or intelligence" means finding succinct representations for our sense data. Solving NP-complete problems would give us optimally compressed representation. For example, you could find a concise mathematical model that would explain stock market data, or Shakespeare's plays or Beethoven's symphonies, provided only that such a model exists to be found. "We're talking about taking some astronomically large space of possibilities, and if there is a solution in that space then finding it efficiently—that's what we mean if NP-complete problems are in P," says Aaronson.

Therefore, an exponential speedup for NP-complete problems on quantum computers would be a big deal. Aaronson thinks it's not likely, but small changes to physical laws could make that happen. In 1998, Daniel S. Abrams and Seth Lloyd showed that "small nonlinearities in time evolution" of quantum states—that is, changing our conception of the time-space continuum—could lead to the design of quantum computers that would solve NP-complete problems in polynomial time. In other words, a change in physical law would be sufficient to give quantum computers this godlike power.

Aaronson has studied many proposals for solving NP-complete problems by using physical mechanisms from soap bubbles to closed timeline curves. In a paper entitled "NP-Complete Problems and Physical Reality," he surveyed the major proposals and even did some experimentation of his own. One paper cited in Aaronson's review suggested that soap bubbles could be used as an analog computer to solve an NP-complete problem (the Steiner tree problem), since soap bubbles tend to settle into shapes having minimal energy. Aaronson built a soap bubble machine, but it stopped finding the optimal solution after only five nodes—soap bubbles won't help a traveling salesman anytime soon. Computing with physical entities (like soap bubbles or DNA) by itself doesn't solve large-scale NP-complete problems. Aaronson the theorist knows the enormously sophisticated literature on NP completeness, including results that give suggestive evidence—though no proofs—that P and NP really are different on digital computers. He also knows the evidence that for a wide variety of problems, Grover's quadratic speedup is the best one can do on quantum computers using accepted physical laws. Finally, there are the strange results showing that if physical laws change, then quantum computers can solve NP-complete problems in polynomial time.

In the absence of a proof that NP-complete problems are fundamentally more difficult than P problems, even for quantum computers, Aaronson has decided to think like a physicist. The standard for raising an assertion to the status of a physical law is lower than that for establishing the truth of a mathematical theorem. After all, there cannot be a proof that entropy never decreases or that no body with positive mass can go faster than the speed of light. These "laws" are consistent with all known observations, and there is a mathematical theory that makes correct predictions assuming they are true, but that is not a proof. Does the inherent difficulty of NP-complete problems have the same property of being a *physical* law but unprovable?

Aaronson notes that many of the deepest principles in physics are impossibility statements. For example, there's no such thing as a perpetual motion machine. "What intrigues me is that there is a two-way relationship between these principles and proposed counterexamples to them," he says. In one direction, a physical law gains credence every time an attempt at invalidating the law fails. In the other direction, sometimes the physical law can lead to discoveries.

Jacob Bekenstein discovered *black-hole entropy*—the theory that black holes would move toward greater disorder like the rest of the universe, subject to the second law of thermodynamics. He arrived at this theory by taking seriously the impossibility of entropy decrease. So, should this *NP hardness assumption*—loosely speaking, the idea that NP-complete problems are intractable in the physical world—eventually be seen as a principle of physics?

In Aaronson's view, the answer ought to depend on (1) whether there is good evidence for the assumption, and (2) whether accepting the assumption places interesting constraints on new physical theories. Aaronson believes that valiant attempts at solving NP-complete problems have foundered for fundamental physical reasons, like not being able to achieve the energy needed to accelerate to relativistic speed comparable to the speed of light. As a result, the hardness assumption has resisted many serious attempts at overcoming it.

The assumption that NP-complete problems are truly intractable (that they truly require exponential time to solve) produces predictions about nature. For example, quantum computers can solve NP-complete problems in polynomial time if it is possible for time travel into the past to affect the present. If we make the assumption that NP-complete problems are intractable, then *Back to the Future*–style time travel cannot happen. Wresting physical predictions from an assumption about NP-complete problems may seem strange, but mathematical theories have predicted physical results since Newton. Will making this assumption help us find a theory of everything? Aaronson has the nerve and talent to tackle that question, but for now, he's not saying.

EPILOGUE

▪ ▪ ▪ ▪

In 1959 the physicist Richard Feynman* gave a speech at the American Physical Society entitled "There's Plenty of Room at the Bottom: An Invitation to Enter a New Field of Physics." The event took place in the stone age of modern biology. DNA had just been characterized a few years earlier. The relationship between RNA and amino acids was still unknown. Yet here is Feynman talking about some possibilities:

> Biology is not simply writing information; it is doing something about it. A biological system can be exceedingly small. Many of the cells are very tiny, but they are very active; they manufacture various substances; they walk around; they wiggle; and they do all kinds of marvelous things—all on a very small scale. Also, they store information. Consider the possibility that we too can make a thing very small which

* Two scientists in this book, Ned Seeman and Paul Rothemund, have won the Feynman Prize in Nanotechnology.

does what we want—that we can manufacture an object that maneuvers at that level.

In a few sentences, Feynman suggested cells that could compute. Other "rooms at the bottom," according to Feynman, included wires having 10-atom diameters and the ability to manufacture by manipulating individual atoms—both have recently become possible. At those nano levels, Feynman anticipated that quantum effects would take over—a prospect he found appealing.

One can imagine how some people in the audience might have scoffed at this prospect. But it's happening. The question is, Where might this all lead? Let's start with the technology.

Computing will escape its self-imposed digital electronic prison. Industrial design schools discovered the principle "form follows function" in the mid twentieth century. Future computing may discover an equivalent principle: "form follows nature." If you want a device that will repair skin, bones, or arteries, it makes far more sense to build the device out of lifestuff (DNA, viruses, or cells) than to build it out of electronics. If you want to understand behavior at the quantum level, it makes more sense to pose questions on a quantum computer than on a digital one. Similarly, if you have questions about fluids, materials, or physical forces, you might do better "measuring" the response of an analogous physical system rather than calculating the result of a differential equation.

Computers will be expected to fend for themselves. Computing at the cellular or DNA level must be robust and allow thousands of cell deaths and accidents. Human society has already learned such robustness. Charles de Gaulle encapsulated this notion with characteristic cynicism when he said, "The cemeteries of the world are filled with indispensable men."

When spacecraft travel many light-days away from Earth, their computers will have to survive on their own because Earthlings can be of only limited help at that distance. Even very terrestrial trading systems will have to survive with little human intervention in the rough-and-tumble of chaotic markets. The result will be a new notion of *smart*: survive, adapt, and do your job well.

Coordination will multiply benefits. Imagine a future in which the cellular computer monitoring your arteries detects a slight abnormality in your heartbeat. The cellular computer communicates to an outside analog computer that embodies a model of your heart and determines that your arteries might need a cleaning. A digital computer takes this output and orders special cleaning bacteria in capsule form to be delivered to your home for you to ingest over the next two weeks. The cellular computer monitors progress to determine whether you improve as anticipated. None of these computing devices could save you individually, but together they could.

So far, we've spoken like technologists. Now let's talk like citizens. A therapeutic cell that calls for help could easily turn into a spy. Fast computers could make better weapons. Is this new technology something to fear? We don't think so. Remember that a few hundred 1950s-era hydrogen bombs could have killed most of humanity. As for surveillance, pencil-sized cameras have existed for a long time, not to mention pacemakers that timestamp moments of excitement. The computing-meets-nature technologies described here are inefficient for doing evil, but they present great opportunities for doing good.

If we are going to be a space-faring civilization, we will need help. A complex spacecraft having thousands or millions of computers cannot afford to be as unreliable as our personal computers. Similarly, if we are going to extend our life spans, we will need therapies that adapt to our needs. Finally, if we are to survive as a species, we will have to make sure that large-scale engineering artifacts—nuclear power plants and strategic missiles—don't destroy us and the planet. In a future world in which *smart* means "adapt and do your job well," smart machines will be our friends, our prosthetics, and our fellow explorers.

NATURAL COMPUTING TIME LINE

▪ ▪ ▪ ▪

1673 Gottfried Wilhelm Leibniz invents a machine to do multiplication.

1805 Joseph-Marie Jacquard makes a weaving device based on holes punched in cards.

1821 Charles Babbage designs his analytical engine to do calculations.

1859 Charles Darwin proposes the theory of natural selection in his *On the Origin of Species.*

1864 Herbert Spencer publishes *Principles of Biology,* which applies key concepts of evolution to the social sciences.

1866 Gregor Mendel publishes his paper "Experiments in Plant Hybridization," which becomes the basis of modern genetics.

1921 The word *robot* is coined by Czech playwright Karel Čapek in his production of *R.U.R.* (*Rossum's Universal Robots*).

1922 Niels Bohr wins a Nobel Prize for work on quantum physics.

1927 Vannevar Bush and his MIT team begin the design of a differential analyzer, a sophisticated mechanical analog computer to solve ordinary differential equations.

1931 Kurt Gödel proves there are statements in mathematics that are true but can't be proved.

1932 Alonzo Church invents lambda calculus, which helps define programming languages.

1935 Alan Turing defines a computation model and shows that some problems are impossible for a computer to solve.

1941 Karl Zuse builds an electromechanical computer in Germany that can store its own programs.

1942 The first self-sustaining atomic reaction is achieved under Stagg Field at the University of Chicago.

1945 J. Presper Eckert and John William Mauchly develop the idea of storing programs and data in the same memory space as part of the ENIAC project at the Moore School of Engineering at the University of Pennsylvania. John von Neumann consults for the companion EDVAC project and writes up notes describing the architecture, which thereafter becomes known as a *von Neumann architecture.*

1947 John von Neumann develops cellular automata as a model for self-replicating machines.

1948 The Hixon Symposium at Caltech includes papers on self-replicating machines.

1948 Norbert Wiener publishes *Cybernetics: or Control and Communication in the Animal and the Machine*, a foundational work in systems theory.

1953 James Watson and Francis Crick discover the DNA double helix, using diffraction images taken by Rosalind Franklin.

1954 Nils Barricelli performs a computer simulation of evolution.

1955 Jonas Salk begins the first immunizations with a killed poliovirus vaccine.

1956 John McCarthy introduces the term *artificial intelligence* at the Dartmouth Conference.

1957 The Soviet Union launches *Sputnik*, the first artificial Earth-orbiting satellite.

1958 Canada and the United States establish the North American Aerospace Defense Command (NORAD) in response to the threat of Soviet bombers.

1959 Richard Feynman gives a speech at the annual meeting of the American Physical Society—"There's Plenty of Room at the Bottom: An Invitation to Enter a New Field of Physics"—mapping out the challenges of modern nanotechnology.

1962 The world's first working robot debuts on the assembly line at General Motors.

1962 Albert Sabin's attenuated poliovirus, orally administered, is released to the public.

1965 Intel cofounder Gordon E. Moore publishes a paper in which he observes that the number of transistors that can be placed on an integrated circuit has increased exponentially, doubling every two years.

1967 ILLIAC IV, the first massively parallel computer is built at the University of Illinois.

1975 *Adaptation in Natural and Artificial Systems*, by John Holland, synthesizes the principles of modern genetic algorithms.

1976 *Viking 1*, the first Mars lander, sends back pictures of the surface of the red planet.

1983 Ned Seeman builds a stable nonlinear DNA structure.

1986 Honda creates an autonomous walking robot, Asimo, named in honor of sci-fi guru Isaac Asimov.

1993 Lee Rubel of the University of Illinois publishes a paper on the extended analog computer, which overcomes the limitations of Shannon's general-purpose analog computer.

1993 Leonard Adleman uses DNA to compute an optimal solution to a close cousin of the traveling-salesman problem.

1994 Peter Shor shows that quantum computers can, in principle, factor numbers exponentially faster than digital computers can, possibly paving the way to cracking a whole class of codes.

1996 Hal Abelson, Gerry Sussman, and Tom Knight coin the term *amorphous computing* for the harnessing of multitudes of asynchronous parallel processors with only local interaction.

1996 Lov Grover shows that quantum computers can, in principle, search for a match of *n* items in *sqrt*(*n*) time, whereas digital computers cannot.

1996 The *Mars Pathfinder* is launched.

1997 Two exploration robots from the *Pathfinder* become the first geologists on Mars.

2004 IBM's Blue Gene becomes the world's fastest computer.

2009 David Shaw unveils Anton, a machine with 512 processors that simulates molecular dynamics.

ACKNOWLEDGMENTS

Science can be only partly explained by words. To help us translate complex ideas into accessible images, we turned to the young and talented Aidan Daly, who drew each chapter's witty opening cartoons and lively and insightful diagrams. We have also benefited from the comments of two excellent readers: Aaron Apple and Glenn Butterfoss.

Numerous friends and family encouraged us along the way, including Claude Chereau, Tom and Sophie Kent, Ann Metcalfe, Carla Paganelli, and Richard Solomon.

W. W. Norton is an author's dream. Each stage of the publishing process was handled gently by consummate professionals who strive to put out the best book possible. We wish to thank our editor, Brendan Curry, for his good sense and steadfastness; our fastidious but unflappable copy editor, Stephanie Hiebert, who patiently considered our changes to her changes; production, Devon Zahn; and design, Brian Mulligan. All were a pleasure to work with.

Finally, special thanks to the scientists profiled in *Natural Computing*. They were extraordinarily generous with their time, ideas, and perspectives about the future.

REFERENCES

■ ■ ■ ■

Chapter 1: RODNEY BROOKS

Brooks, Rodney A. "Elephants Don't Play Chess." *Robots and Autonomous Systems* 6 (1990): 3–15.

Brooks, Rodney A., and Anita M. Flynn. "Fast, Cheap and Out of Control: A Robot Invasion of the Solar System." *Journal of the British Interplanetary Society*, 1989, 478–85.

Spenko, M. J., G. C. Haynes, J. A. Saunders, A. A. Rizzi, R. J. Full, and D. E. Koditschek. "Biologically Inspired Climbing with a Hexapedal Robot." *Journal of Field Robotics*, April 2008, 223–42.

Chapter 2: GLENN REEVES AND ADRIAN STOICA

Bell, Jim. *Postcards from Mars: The First Photographer on the Red Planet*. New York: Dutton, 2006.

Koza, John R., F. H. Bennett III, D. Andre, and M. A. Keane. *Genetic Programming III—Darwinian Invention and Problem Solving.* San Francisco: Morgan Kaufman, 1999.

Reeves, Glenn. "What Really Happened on Mars?" *Risks Digest* 19, no. 49 (1997).

Reeves, G. E., and J. F. Snyder. "An Overview of the Mars Exploration Rovers' Flight Software." Paper presented at the IEEE International Conference on Systems, Man, and Cybernetics, Waikoloa, HI, October 10–12, 2005.

Stoica, Adrian, Ricardo Zebulum, Didier Keymeulen, Rajeshuni Ramesham, Joseph Neff, and Srinivas Katkoori. "Temperature-Adaptive Circuits on Reconfigurable Analog Arrays." Paper presented at the first NASA/ESA Conference on Adaptive Hardware and Systems, Istanbul, Turkey, June 15–18, 2006.

Stoica, Adrian, Ricardo Zebulum, Didier Keymeulen, Raoul Tawel, Taher Daud, and Anil Thakoor. "Reconfigurable VLSI Architectures for Evolvable Hardware: From Experimental Field Programmable Transistor Arrays to Evolution-Oriented Chips." *IEEE Transactions on Very Large Scale Integration Systems* 9, no. 1 (2001): 227–32.

Chapter 3: **LOUIS QUALLS**

Mason, Lee, David Poston, and Louis Qualls. "System Concepts for Affordable Fission Surface Power." Paper presented at the Space Technology and Applications International Forum (STAIF–2008), sponsored by the Institute for Space and Nuclear Power Studies at the University of New Mexico, Albuquerque, NM, February 10–14, 2008.

Chapter 5: **NANCY LEVESON**

Leveson, Nancy. *System Safety Engineering: Back to the Future* (a work in progress available on Leveson's website, http://sunnyday.mit.edu/book2.html).

Chapter 6: **NED SEEMAN**

Seeman, Nadrian C. "Nucleic-Acid Junctions and Lattices." *Journal of Theoretical Biology* 99, no. 2 (1982): 237–47.

Zhang, Y., and Nadrian C. Seeman. "The Construction of a DNA Truncated Octahedron." *Journal of the American Chemical Society* 116 (1994): 1661–69.

Chapter 7: **PAUL ROTHEMUND**

Adleman, Leonard M. "Computing with DNA." *Scientific American*, August 1998, 54–61.

Adleman, Leonard M. "Molecular Computation of Solutions to Combinatorial Problems." *Science* 266 (1994): 1021–24.

Amos, Martyn. *Genesis Machines: The New Science of Biocomputing.* New York: Overlook Press, 2008.

Benenson, Y., B. Gil, U. Ben-Dor, R. Adar, and E. Shapiro. "An Autonomous Molecular Computer for Logical Control of Gene Expression." *Nature* 429 (2004): 423–29.

Charles H. Bennett. "The Thermodynamics of Computation—A Review." *International Journal of Theoretical Physics* 21 (1982): 905–40.

Rothemund, Paul W. K. "A DNA and Restriction Enzyme Implementation of Turing Machines." In *DNA Based Computers: Proceedings of a DIMACS Workshop, April 4, 1995, Princeton University* (DIMACS Series in Discrete Mathematics and Theoretical Computer Science 27), edited by Richard J. Lipton and E. B. Baum, 75–119. Providence, RI: American Mathematical Society, 1996.

Winfree, Erik. "On the Computational Power of DNA Annealing and Ligation." In *DNA Based Computers: Proceedings of a DIMACS Workshop, April 4, 1995, Princeton University* (DIMACS Series in Discrete Mathematics and Theoretical Computer Science 27), edited by Richard J. Lipton and

E. B. Baum, 199–221. Providence, RI: American Mathematical Society, 1996.

Winfree, Erik, Furong Liu, Lisa A. Wenzler, and Nadrian C. Seeman. "Design and Self-Assembly of Two-Dimensional DNA Crystals." *Nature* 394 (1998): 539–44.

Yurke, Bernard, Andrew J. Turberfield, Allen P. Mills Jr., Friedrich C. Simmel, and Jennifer L. Neumann. "A DNA-Fuelled Molecular Machine Made of DNA." *Nature* 406 (2000): 605–8.

Chapter 8: **STEVE SKIENA**

Coleman, J. R., D. Papamichail, S. Skiena, B. Futcher, E. Wimmer, and S. Mueller. "Virus Attenuation by Genome-Scale Changes in Codon Pair Bias." *Science* 320 (2008): 1784–87.

Skiena, Steve. *Calculated Bets: Computers, Gambling, and Mathematical Modeling to Win.* Cambridge: Cambridge University Press, 2001.

Chapter 9: **GERALD SUSSMAN**

Abelson, Harold, Don Allen, Daniel Coore, Chris Hanson, George Homsy, Thomas F. Knight Jr., Radhika Nagpal, Eric Rauch, Gerald Jay Sussman, and Ron Weiss. "Amorphous Computing." *Communications of the ACM*, May 2000, 74–82.

Weiss, Ron, Thomas F. Knight Jr., and Gerald Sussman. "Genetic Process Engineering." Chapter 4 in *Cellular Computing*, edited by Martyn Amos. Oxford: Oxford University Press, 2004.

Chapter 10: **RADHIKA NAGPAL**

Coore, Daniel, Radhika Nagpal, and Ron Weiss. "Paradigms for Structure in an Amorphous Computer." A. I. Memo no. 1614. Cambridge, MA: MIT Artificial Intelligence Laboratory, 1997.

Nagpal, Radhika. "Programmable Self-Assembly: Constructing Global Shape Using Biologically-Inspired Local Interactions and Origami Mathematics." PhD diss., MIT, 2001.

Nüsslein-Volhard, C. "Gradients That Organize Embryo Development." *Scientific American*, August 1990, 54–55, 58–61.

Chapter 11: **MONTY DENNEAU**

Almási, George, Călin Caşcaval, José G. Castaños, Monty Denneau, Derek Lieber, José E. Moreira, and Henry S. Warren Jr. "Dissecting Cyclops: A Detailed Analysis of a Multithreaded Architecture." *SIGARCH Computer Architecture News* 31, no. 1 (2003): 26–48.

Beetem, John, Monty Denneau, and Don Weingarten. "The GF11 Supercomputer." In *Proceedings of the 12th Annual International Symposium on Computer Architecture*, 108–15. Los Alamitos, CA: IEEE Computer Society Press, 1985.

Chapter 12: **DAVID SHAW**

Bowers, Kevin J., Ron O. Dror, and David E. Shaw. "Overview of Neutral Territory Methods for the Parallel Evaluation of Pairwise Interactions." *Journal of Physics: Conference Series* 16 (2005): 300–304.

Shaw, David E., Martin M. Deneroff, Ron O. Dror, Jeffrey S. Kuskin, Richard H. Larson, John K. Salmon, Cliff Young, Brannon Batson, Kevin J. Bowers, Jack C. Chao, Michael P. Eastwood, Joseph Gagliardo, J. P. Grossman, C. Richard Ho, Douglas J. Ierardi, Istvan Kolossvary, John L. Klepeis, Timothy Layman, Christine McLeavey, Mark A. Moraes, Rolf Mueller, Edward C. Priest, Yibing Shan, Jochen Spengler, Michael Theobald, Brian Towles, and Stanley C. Wang. "Anton, a Special-Purpose Machine for Molecular Dynamics Simulation." *Communications of the ACM* 51, no. 7 (2008): 91–97.

Chapter 13: **JONATHAN MILLS**

Karplus, W. *Analog Simulation.* New York: McGraw-Hill, 1958.

Mills, J. "The Nature of the Extended Analog Computer." *Physica D* 237 (2008): 1235–56.

Mills, J. W., M. Parker, B. Himebaugh, C. Shue, B. Kopecky, and C. Weilemann. "'Empty Space' Computes: The Evolution of an Unconventional Supercomputer." In *Proceedings of the 3rd Conference on Computing Frontiers,* 115–26. New York: Association for Computing Machinery, 2006.

Rubel, L. A. "The Extended Analog Computer." *Advances in Applied Mathematics* 14 (1993): 39–50.

Chapter 14: **SCOTT AARONSON**

Deutsch, David. *The Fabric of Reality: The Science of Parallel Universes and Its Implications.* New York: Penguin, 1998.

Grover, Lov. "A Fast Quantum Mechanical Algorithm for Database Search." In *Proceedings of the Twenty-Eighth Annual ACM Symposium on Theory of Computing,* 212–19. New York: ACM, 1996.

Shor, Peter W. "Polynomial-Time Algorithms for Prime Factorization and Discrete Logarithms on a Quantum Computer." *SIAM Journal on Computing* 26 (1997): 1484–1509.

INDEX

Page numbers in *italics* refer to illustrations.
Names in **boldface** refer to the principal subject of a chapter.